AS CORE MATHS 1, 600+ CHALLENGING QUESTIONS FOR A* STUDENTS

M DON

ISBN: **10: 1500357553**
ISBN-13: **978-1500357559**

DEDICATION

To my older brother, R. George, who taught me to how do fractions when I was a little kid! And also to my beloved teachers S.Cooray and K.S Chandranath, whom I was lucky enough to meet in this life time.

ACKNOWLEDGMENTS

I would like to thank all my students for their great input while preparing this book and for all the knowledge and the jokes we have shared among us.

For the Students:

About the book:

The main purpose of this book is to provide with some ample amount of C 1 questions with challenging nature for serious and sincere students who are aiming to get excellent grades in core math 1 and excel in mathematics in general. Though the questions are mainly focused on core 1, they can be very useful as a preparation for other core modules as well.

Some questions in this book might need the knowledge and understand of more than one chapter, for an example a question in chapter 4 (Sketching curves) might need the knowledge of chapter 2 (sketching curves) and differentiation (chapter 6). This has done deliberately in order to give the student maximum exposure to the materials in different contexts. This will improve one's linking ability, which is quite important when it comes to examinations.

This book tests ones fundamental knowledge in core math as well as thinking and linking skills which is required for the exams.

How to use this book:

It's recommended to try these questions alongside (or after) the questions students will find in standard school texts. So the student will get the opportunity to get the best of both worlds! However this is not essential as this book also contains a considerable amount of such questions in the beginning of each chapter. Or even one can consider trying these questions with past exam paper questions to a greater effect.

To Contact the Author

Any student can contact the author via" **core_mathematics@ yahoo.co.uk** "or by phone **07412358153** (preferred through" **whatsapp**" messenger) for any answers or guidance. And also any suggestions and comments about this book will be much appreciated.

CONTENTS

Simplify the following expressions: 1A

1. $12y^{-3} \div 4y^{-5}$ **2.** $(3z^2)^{-2} \div z^4$ **3.** $4yx^2 \div 8y^3$ **4.** $A^{-3}B^2 \div (A^{-1}B)^{-2}$

5. $\dfrac{4x^{-3}}{(2^3x)^{-2}}$ **6.** $\dfrac{(xy)^{-2}}{2^{-2}(yx^{-2})^{-1}}$ **7.** $\left(\dfrac{yx^{-3}}{xy^{-2}}\right)^{-1}$ **8.** $\left(\dfrac{4x^{-2}}{81z^2}\right)^{\frac{1}{2}}$

9. $\left(\dfrac{9z^4}{16x^{-4}}\right)^{-0.5}$ **10.** $\sqrt{\dfrac{36y}{25y^3x^{-2}}}$ **11.** $\sqrt{\dfrac{0.5x^{-2}}{18y^{-4}}}$ **12.** $\sqrt[3]{\dfrac{0.25x^{\frac{2}{3}}}{2}}$

13. $\sqrt{\dfrac{0.2y^{-4}}{5}}$ **14.** $\sqrt[4]{\dfrac{4z^{-4}}{0.25}}$ **15.** $\sqrt{\dfrac{x^2+2xy+y^2}{25}}$

16. $\left(\dfrac{36}{z^2+2xz+x^2}\right)^{-0.5}$ **17.** $x+\dfrac{y^2-xy}{x-y}$ **18.** $y+\dfrac{x^2+xy}{x+y}$

19. $\sqrt{\dfrac{x^2+2xy+y^2}{x^2-2xy+y^2}}$ **20.** $\left(\dfrac{x^2+2xy+y^2}{x^2-2xy+y^2}\right)^{\frac{-3}{2}}$ **21.** $(-1)^{\frac{2}{3}}$

22. $(-8)^{\frac{2}{3}}$ **23.** $\left(\dfrac{0.2y^{-6}}{5}\right)^{\frac{1}{2}}$

Factorise completely: **1B**

1. $12x^2y + 8xy^2$ **2.** $25x^2 - 100$ **3.** $x^3 - xy^2$ **4.** $36A^2 - 81$ **5.** $7t^2 - 28$

6. $8A^2 - 50B^2$ **7.** abc + bcd + a + d **8.** $pqr + qrs + p^2 - s^2$ **9.** abc − bcd − a + d

10. $p^2qr - qrs^2 + p + s$ **11.** $9a^2b^2 - 25c^2d^2$ **12.** $z^4 - 25z^2$ **13.** $49t^4 - 9t^2$

Factorise completely: **1C**

1. $x^4 - y^4$ **2.** $16A^4 - 81B^4$ **3.** $x^2 + 2x + 1$ **4.** $x^2 - 8x + 12$ **5.** $4p^2 + 8p + 4$

6. $15x^2 + 42x + 9$ **7.** $6x^2 + 11x - 7$ **8.** $35x^2 + 8x - 3$ **9.** $x^4 - y^4 + 4x - 4y$

10. $p^4 + 2p^2 + 1$

1D

1. **a** . Factorise , $p^2 + 5p + 6$,

   b. Hence or otherwise factorise $p^2 + 5bp + 6b^2$

   c. Given that $p^2 + 5bp + 6b^2 = 0$, find two relationships between p and b.

2. a. factorise , $x^2 + 5x + 6$

   b. Hence or otherwise factorise $(x + p)^2 + 5(x + p) + 6$

   c. Given $(x + p)^2 + 5(x + p) + 6 = 0$, Write two relationships between p and x.

3. a. factorise , $6t^2 + 11t - 7$
   b. hence factorise , $6t^2 + 11bt - 7b^2$ and $6p^2t^2 + 11pt - 7$

4. factorise , $x^2 + 5x + 6$ hence evaluate $(997)^2 + 5 \times 997 + 6$

5. factorise , $x^2 - x - 6$ hence evaluate $(1003)^2 - 1009$

6. f(x) $= ax^2 + bx + c$, given that $b = a + c$,
   a. Factorise f(x) completely
   b. Hence factorise **(i)** $95x^2 + 25x - 70$ **(ii)** $px^2 + x + 1 - p$

7. If $p x^2 + qx + r \equiv (1+x)(px + r)$

    a. Show that $p + r = q$

    b. Hence factorise **(i)** $-8 x^2 - 3x + 5$ **(ii)** $(a - b)x^2 + 3ax + 2a + b$

8. $f(x) = x^4 + 5x^2 + 6$,

    a. By substituting $x^2 = T$ factorise $f(x)$

    b. Hence factorise $f(x - 3)$.

9. $f(x) = 1 + \dfrac{2}{x} + \dfrac{1}{x^2}$

    a. by substituting $x^{-1} = T$ factorise $f(x)$.

    b. hence factorise $1 + \dfrac{2}{x-1} + \dfrac{1}{(x-1)^2}$

10. By substituting $p - 1 = x$, factorise $f(p) = 4 - \dfrac{5}{p-1} - \dfrac{9}{(p-1)^2}$

11. Factorise $x^4 - 2x^2 - 3$, hence simply the expression $\dfrac{x^2 - 3}{x^4 - 2x^2 - 3}$

12.

    a. Factorise $2p^3 + 2p^2 - 9p - 9$

    b. Hence simplify $\dfrac{2p^3 + 2p^2 - 9p - 9}{2p^2 - 9}$

    c. using part (b) evaluate $\sqrt{\dfrac{2(99)^3 + 2(99)^2 - 9(99) - 9}{2 \times (99)^2 - 9}}$

13. Factorise : $x^2 - y^2$

   b. hence evaluate $\sqrt{36^2 - 4^2}$ giving your answer in simplest surd form .

   c. factorise $(2p - 1)^2 - (3p + 4)^2$

14. by factorising or otherwise show that $(2p - 1)^4 - (2p + 1)^4 \equiv -16p(4p^2 + 1)$

15. Show that $\sqrt{(2p + 2)^4 - (2p - 2)^4} \equiv 8\sqrt{p(p^2 + 1)}$, hence and by substituting a suitable value for p evaluate $\sqrt{6^4 - 2^4}$

16. Show that $\dfrac{x^2 - y^2 + x + y}{x - y + 1} \equiv x + y$, hence evaluate $\dfrac{95^2 - 5^2 + 100}{91}$

17. Factorise $4p^2 - 16q^2$ hence or otherwise show that $\dfrac{4p^2 + 2p + 4q - 16q^2}{2p - 4q + 1} \equiv 2p + 4q$

18. Utilise factorisation to Show that $\dfrac{abc + bcd + a + d}{bc + 1} \equiv a + d$

19. Utilise factorisation to Show that $\dfrac{p + qr - s}{p^2 + pqr + qrs - s^2} \equiv \dfrac{1}{p + s}$

20. Factorise $K x^2 + x + 1 - k$, hence or otherwise simplify the expression $\dfrac{3x + 3}{K x^2 + x + 1 - k}$

21. By utilising the fact $\sqrt{3}^2 = 3$, factorise $(x+2)^2 - 3$.

22. a . Factorise $(p+2)^2 - 4q^2$ **b.** given that $(p+2)^2 - 4q^2 = 0$ find two possible relationships between p and q.

Evaluate :   1E

1. 37^0 **2.**$(-5)^2$ **3.** $(-27)^{\frac{1}{3}}$ **4.** $\left(\frac{7}{4}\right)^{-2}$ **5.** $\left(\frac{64}{27}\right)^{\frac{1}{3}}$ **6.** $\left(1\frac{9}{16}\right)^{\frac{3}{2}}$

7. $\left(1\frac{9}{16}\right)^{\frac{-3}{2}}$ **8.** $\sqrt{\frac{0.5}{18}}$ **9.** $\sqrt[3]{\frac{2}{27}} \times \sqrt[3]{32}$ **10.** $5^{\frac{1}{3}} \times \left(\frac{25}{64}\right)^{\frac{1}{3}}$

11. $3^{\frac{1}{3}} \times \left(\frac{27}{81}\right)^{\frac{1}{3}}$ **12.** $\left(\frac{1}{2}\right)^{\frac{-1}{3}} \times (52)^{\frac{-1}{3}}$ **13.** $(0.2)^{\frac{1}{3}} \times \left(\frac{1}{25}\right)^{\frac{1}{3}}$

14. $3^{\frac{1}{3}} \div \left(\frac{81}{27}\right)^{\frac{1}{3}}$ **15.** $\sqrt[3]{\frac{27}{2}} \div \sqrt[3]{32}$ **16.** $\left(\frac{1}{0.5}\right)^{\frac{-1}{3}} \div (52)^{\frac{-1}{3}}$

17. $\sqrt{\frac{2}{27}} \div \sqrt{\frac{32}{3}}$ **18.** $\sqrt[3]{\frac{1}{12}} \div \sqrt[3]{\frac{16}{3}}$

19. $\sqrt[3]{0.2} \times \sqrt[3]{0.04}$ **20.** $\sqrt[3]{0.5} \times \sqrt[3]{0.54}$ **21.** $\sqrt[3]{2} \times \sqrt[3]{0.08}$

22. $\sqrt{5} + \dfrac{60}{\sqrt{45}} = C\sqrt{5}$, find the value of **C**.

23. $\sqrt{52} - \dfrac{K}{\sqrt{13}} = \sqrt{13}$, find the value of K.

24. Evaluate $(\sqrt{2} - 1)(\sqrt{7} - 3)$

25. Consider the expansion of the expression , (x-3)(x-K)(x+3), coefficient of x^2 is 10. Find the value of K

26. Consider the expansion of the expression , (x-1)(x-K)(x+3), the constant of the expansion is equal to -3. Find the value of K

24. Expand $(\sqrt{2} - \sqrt{3})(1 + \sqrt{2})(\sqrt{2} + \sqrt{3})$

25. $(\sqrt{2} - \sqrt{3})(1 - \sqrt{K})(\sqrt{2} + \sqrt{3}) = a\sqrt{3} + b$, find K , a and b

26. Simplify by rationalising the denominator , $\dfrac{\sqrt{2}}{1 - 2\sqrt{2}}$

27. $\dfrac{3 + 2\sqrt{2}}{1 + \sqrt{2}} = \sqrt{k} + p$ find K and P.

28. Expand and simplify $(1 + \sqrt{3})^2$, hence workout $\dfrac{4 + 2\sqrt{3}}{3 + 3\sqrt{3}}$

29. Expand : $(1 - \sqrt{a})^2$, hence simplify , $\dfrac{1 - a - 2\sqrt{a}}{2 - 2\sqrt{a}}$

30. Given that $2^{x-1} = 16^y$ show that $4y - x + 1 = 0$

31. Given that $2^3 = (2k)^{-3}$ find the value of k.

32. a. Factorise $x^2 - y^2$,

    b. hence or otherwise find k where $(\sqrt{a} + \sqrt{b})^2 - (\sqrt{a} - \sqrt{b})^2 = k\sqrt{ab}$.

    c. evaluate $(\sqrt{7} + \sqrt{3})^2 - (\sqrt{7} - \sqrt{3})^2$

33. expand $(\sqrt{a} + \sqrt{b})^2$, hence by substituting suitable values for a and b show that

$$\sqrt{\frac{3}{10+2\sqrt{21}}} = \frac{\sqrt{3}}{\sqrt{7}+\sqrt{3}}$$

34. expand $(\sqrt{a} - \sqrt{b})^2$, hence by substituting suitable values for a and b show that

$$\left(\frac{3}{4-2\sqrt{21}}\right)^{\frac{1}{2}} = \frac{\sqrt{3}}{\sqrt{7}-\sqrt{3}}$$

35. Expand $(\sqrt{2} - \sqrt{3})^2$, hence show that $\sqrt{\frac{8}{2\sqrt{6}-1}} = \frac{2\sqrt{2}}{\sqrt{2}-\sqrt{3}} = -2\sqrt{6} - 4$

36. Expand $(\sqrt{5} - \sqrt{3})^2$, Expand $(\sqrt{5} - \sqrt{3})^2$, show that $\sqrt{\frac{2-2\sqrt{15}}{5}} = 1 - \sqrt{\frac{3}{5}}$

37. if $(\sqrt{2})^{x-1} = 8^y$ show that $6y - x + 1 = 0$ and solve $(\sqrt{2})^{x-1} = 8^{20}$

8

38. if $2^x = \left(\sqrt{12}\right)^y$, show that $\dfrac{2^{x-y}}{3^{0.5y}} = 1$

39. expand : $\sqrt{2}(1 - \sqrt{3})(1 + \sqrt{3})$

40. expand : $T^{\frac{1}{2}}\left(T^{\frac{3}{2}} - 1\right)\left(T^{\frac{1}{2}} + T^{\frac{-1}{2}}\right)$, Hence evaluate , $4^{\frac{1}{2}}\left(4^{\frac{3}{2}} - 1\right)\left(4^{\frac{1}{2}} + 4^{\frac{-1}{2}}\right)$

41. Workout : $\dfrac{1}{\sqrt{2}-1} - \dfrac{1}{\sqrt{2}+1}$

42. Write the expression $\dfrac{1}{\sqrt{2}-1} + \dfrac{1}{\sqrt{2}+1}$ in the form of $K\sqrt{2}$ where k is an integer.

43. $\dfrac{1+\sqrt{2}}{2\sqrt{2}+3} = \dfrac{1}{\sqrt{p}+q}$, find the values of p and q.

44. $\dfrac{2-\sqrt{3}}{\sqrt{3}+5} = \dfrac{1}{7\sqrt{p}+q}$, find the values of p and q.

45. $\dfrac{\sqrt{5}-2}{1-4\sqrt{5}} = \sqrt{p} + q$, by rationalizing the denominator find p and q.

46. $\dfrac{2\sqrt{2}+3}{1+\sqrt{2}} = \sqrt{p} + q$ by rationalizing the denominator find p and q.

47. Expand $(2-\sqrt{x})^2$, hence or otherwise rationalize the denominator of the expression

$$\frac{2-\sqrt{x}}{4+x-4\sqrt{x}} \quad .$$

48. Expand $(1-\sqrt{x})^2$, hence or otherwise rationalize the denominator of the expression

$$\frac{1-x+2\sqrt{x}}{3-3\sqrt{x}}$$

49. Given that $2^{y^2-1} = 16^{x-1}$ show that $y^2-4x+3 = 0$, solve, $2^{99} - 16^{x-1} = 0$

50. Write the summation of $\frac{1}{\sqrt{2}} + \frac{1}{\sqrt{3}} + \frac{1}{\sqrt{5}}$ with a rational denominator.

1F

Simplify by giving the answers in surd form

1. $\frac{\sqrt{27}}{3}$ **2.** $\sqrt{32}$ **3.** $\sqrt{20} + \sqrt{80}$ **4.** $\sqrt{200} + \sqrt{18} +\sqrt{72}$

5. $3\sqrt{80} - 2\sqrt{20} +5\sqrt{45}$ **6.** $\frac{\sqrt{20} + \sqrt{80}}{\sqrt{5}}$ **7.** $\frac{\sqrt{7}}{\sqrt{28} - \sqrt{56}}$

8. $\frac{\sqrt{3}}{\sqrt{3} + \sqrt{27} - \sqrt{12}}$ **9.** $\frac{\sqrt{28} - \sqrt{112}}{\sqrt{63} - \sqrt{28}}$ **10.** $\frac{\sqrt{27} - \sqrt{12}}{\sqrt{75} + \sqrt{48}}$ **11.** $\sqrt{20x} + \sqrt{80x}$

12. $3y\sqrt{80x} - 2y\sqrt{20x} +5y\sqrt{45x}$ **13.** $\frac{\sqrt{7p}}{\sqrt{28p} - \sqrt{56p}}$ **14.** $\frac{\sqrt{27p} - \sqrt{12p}}{\sqrt{75} + \sqrt{48}}$

15. $\frac{q\sqrt{28} - q\sqrt{112}}{\sqrt{63q} - \sqrt{28q}}$ **16.** $\frac{\sqrt{3x}}{x\sqrt{3} +x\sqrt{27} - x\sqrt{12}}$

2: Quadratic Expressions, Equations and Graphs

Basics: 2A

Choose the expressions which are **quadratic** from the following expressions.

a. $Y = 5x^2 + \frac{3}{4}x + 1$ **b.** $Y = 1 - \frac{3}{4}x^2$ **c.** $Y = (1-2x)^2$ **d.** $Y = x^3 - 5x^2$

e. $Y = \frac{2}{5}x^2$ **f.** $Y = -3x^4 + x$ **g.** $Y = 5x^2 + x + (1-5x)^2$ **I.** $y = \frac{x^3+x+1}{x}$

j. $Y = (1-2x)^2 - 4x^2$ **H.** $y = \frac{x^3+x^2-x}{3x}$

1. What is the difference between " an expression and an equation"? Explain your answer algebraically   and graphically.

2. What do you mean by "solving an   equation"? Explain your answer algebraically and graphically

Solve (find the roots) the following equations: 2B

1. $x^2 = 9$ **2.** $x^2 = 36x$ **3.** $x^2 - x - 6 = 0$ **4.** $(4x - 1)^2 = 16$

5. $(p - 1)^2 = 16\ (p + 1)^2$ **6.** $x^2 - 5x + 6 = 0$ **7.** $11\,q^2 + 17q + 6 = 0$

8. $p + 5 + \dfrac{6}{p} = 0$ **9.** $1 + \dfrac{6}{x^2} + \dfrac{5}{x} = 0$ **10.** $\dfrac{-2}{y} = y + 3$

11. $\dfrac{z+3}{z} = \dfrac{z}{3}$ **12.** $\dfrac{4}{x} = x$ **13.** $K(K-1) + 2K = 0$ **14.** $(x-4)(4x+1) = 0$

15. $(-x+1)(X+1) - 1 = 0$ **16.** $\dfrac{4}{x} = x - 3$ **17.** $1 + \dfrac{6}{(1+x)^2} + \dfrac{5}{x+1} = 0$ **18.** $\dfrac{z+4}{z+1} = \dfrac{z+1}{3}$

19. $p-6 + \dfrac{6}{p+1} = 0$ **20.** $\dfrac{1}{(1+x)^2} = \dfrac{3}{(1-x)^2}$

21. solve for x : $3^{-10x-12} = 9^{x^2}$

22. solve for x : $125^x = 5^{x^2}$

23. Given that $y = (1-x)^{2x} = (1-x)^{x^2-3x+6}$, find the values of x and y.

12

24. Given that $y = (1-p) = (1-p)^{p^2-7}$ find the exact values of p and y.

25. solve for x, $(x+1)^{x^2-4} = 1$

26. by substituting $t^2 = y$ or otherwise solve for t, $t^4 - 5t^2 + 6 = 0$

27. by a suitable substitution solve for q, $11\,q^4 + 17\,q^2 + 6 = 0$,

28. by substituting $2t-1 = y$ or otherwise solve for t, $(2t-1)^2 = 5(2t-1) + 6$

29. solve for p, $7(2p+3)^2 - (2p+3) - 6 = 0$

30. solve for x, $(x^2 - 9)(x^2 + 1) = 0$

31. solve for x, $\dfrac{x^2+1}{x^2} = \dfrac{2}{1-x^2}$, hence deduce the roots of $\dfrac{(x+5)^2+1}{(x+5)^2} = \dfrac{2}{1-(x+5)^2}$, give all the answers in surd form.

Find the exact answers by completing the square: 2C

1. $x^2 + 9x + 2 = 0$ **2.** $-x^2 + 7x + 2 = 0$ **3.** $3x^2 + 7x + 2 = 0$

4. $-2x^2 + 7x + 2 = 0$ **5.** $15 = x^2 + 9x$ **6.** $x(x+1) = 6$

7. $\frac{3}{4}x^2 + 9x + 2 = 0$

8. $-\frac{2}{5}x^2 + 7x + 2 = 0$ **9.** $-x^2 + 4x + \frac{2}{3} = 0$ **10.** $-2x^2 + \frac{1}{3}x + 2 = 0$

11. by considering factorization or otherwise solve for x in terms of p, where p is a real number .

$px^2 + 2x + 2 - p = 0$

12. . by considering factorization or otherwise solve for x in terms of p, where p is a real number .

$x^2 + (1 + p)x + p = 0$

13. by considering factorization or otherwise solve for x in terms of p, where p is a real number .

$2x^2 - px - p^2 = 0$, hence find the roots of $2x^2 - 9x - 81 = 0$

14. A quadratic equation has roots x = -3 and x = 2 , and crosses the Y axis at (0,-3), find the equation of the quadratic equation .

15. A quadratic equation has roots x = -1 and x = 2 , and pass through (1,1) , find the equation of the quadratic equation

16. The roots of the equation $x^2 + bx + c = 0$ given by α and β . by completing the square or otherwise

Show that α + β = -b , αβ = c.

Hence or otherwise deduce the **sum** and the **product** of the roots of the equation given by $x^2 - 4x + 1 = 0$

17. The roots of the equation $ax^2 + x + 1 = 0$ given by α and β . by completing the square or otherwise

Show that $\alpha + \beta = -\dfrac{1}{a}$, $\alpha\beta = \dfrac{1}{a}$.

Hence or otherwise deduce the **sum** and the **product** of the roots of the equation given by

$-3x^2 + x + 1 = 0$

18.   The roots of the equation $x^2 + bx + c = 0$ given by α and β . by completing the square or otherwise

    a. Show that $\alpha + \beta = -b$, $\alpha\beta = c$.

    b) show that $\alpha^2 + \beta^2 = (\alpha + \beta)^2 - 2\alpha\beta$

    c) Hence Write an expression for $\alpha^2 + \beta^2$ in terms of b and c.

    d) deduce the **sum of the squares** of the roots of the equation given by $x^2 - 5x + 7 = 0$

19. Given that $x^2 + 4x + 1 \equiv (x + p)^2 + r$

    a. find p and r

    b. hence find the exact roots of $x^2 + 4x + 1 = 0$

20. Given that $5x^2 + 5x - 1 \equiv p(x + q)^2 + r$

    a. find p, q and r

    b. hence find the exact roots of $5x^2 + 5x - 1 = 0$

21. given that , $\dfrac{x-1}{x+1} + \dfrac{2x}{x+2} \equiv \dfrac{P(x+q)^2 +r}{(x+2)(x+1)}$

    a. find p, q and r

    b. hence find the roots of $\dfrac{x-1}{x+1} + \dfrac{2x}{x+2} = 0$, give the answers in exact form.

22. Given that $Y = x^4 - 5x^2 + 6 \equiv p(x^2 + q)^2 + r$

    a. find p, q and r

    b. find the exact roots of $x^4 - 5x^2 + 6 = 0$

    c. find the exact roots of $x^4 - 5x^2 + 6 = 0$

23. State the Line of symmetry and the coordinates of the minimum or maximum point of

    $Y = (x - 1)^2 + 5$

24. State the Line of symmetry and the coordinates of the minimum or maximum point of

    $Y = -(3x - 1)^2 - 5$

25. State the Line of symmetry and the coordinates of the minimum or maximum point of

    $Y = -4(x + 1)^2 + 3$

26. State the Line of symmetry and the coordinates of the minimum or maximum point of

    $Y = -3x^2 + x + 1$

27. State the Line of symmetry and the coordinates of the minimum or maximum point of

$Y = 2x^2 + x + 1$

28. sketch $Y = (x - 1)^2 + 6$ and $Y = 6$ on the same grid, hence deduce the coordinates where

$Y = (x - 1)^2 + 6$ and $Y = 6$ touches each other.

29. Justify that $Y = (x - 1)^2 + 6$ and $Y = 3$ do not intercept

30. $Y = p$ is a tangent to $Y = (x - 1)^2 - 3$. given that p is a constant. deduce the value of p.
Hence write the coordinates where $Y = p$ touches $Y = (x - 1)^2 - 3$.

31. $Y = P$ is a tangent to $Y = 2x^2 + x + 1$. Given that p is a constant. Deduce the value of p.
Hence write the coordinates where $Y = p$ touches $Y = 2x^2 + x + 1$

32. Minimum point of the curve $Y = x^2 + bx + c$ is given by (3,-4). Find the values of b and c.

33 Maximum point of the curve $Y = -2x^2 + bx + c$ is given by (3,-4). Find the values of b and c

34. The diagram shows a quadratic graph,

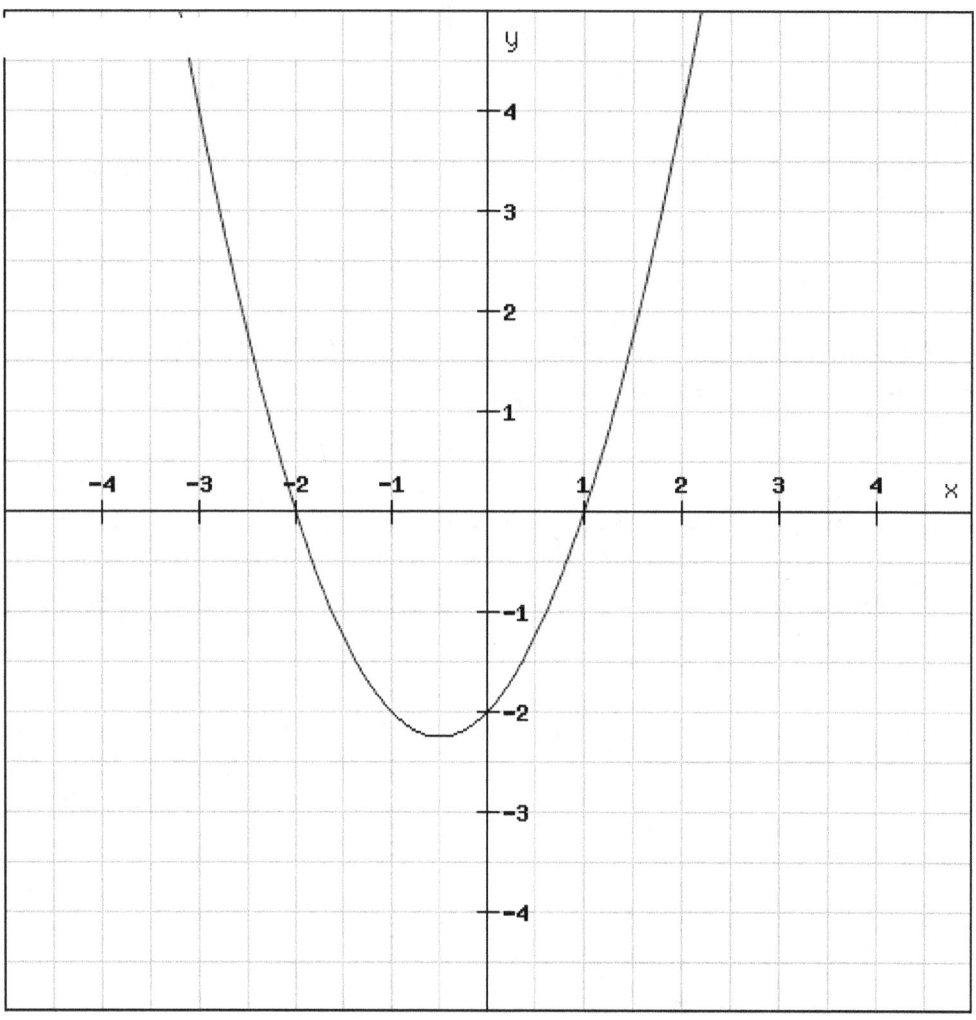

The equation of the graph given by $y = K(X+p)(X+q)$,

a. Use the graph to find the vales of K, p and q.

b. Draw the line of symmetry .

c. Write the equation of the line of symmetry.

d. Hence or otherwise find the minimum value of Y.

e. Hence or otherwise Sketch $y = k(X+p+2)(X+q+2)$ on the same grid

35. The quadratic graph is given by the expression $y = k(x-P)(x-5)$. the minimum point of the graph is (3,-8).

a. Find the values of K, and p

b. Find the Y intercept of the graph

c. Sketch $Y = (x-P)(x-5)$ on the same grid, clearly showing the coordinates of x intercepts , Y intercept and the minimum point.

36. The graph of the quadratic function given by $Y = K(2x-a)(3x-b)$, intercept X axis at (2,0) and (3,0)

a. State the value of a and b

b. Given that the Y intercept of the graph is (0, -3), expand and complete the square of $K(2x-a)(3x-b)$

c. Hence find the Maximum value of Y.

37. quadratic graph given by $Y = (p-X)(X-q)$. And the graph cuts the Y axis at (0, -10) . The line Y = 3 intercepts the graph at (2,3) and (7,3) .

a. State the value of p + q

b. Hence write the equation of the line of symmetry.

c. Show that pq = -10

d. Find the values of p and q , Hence find the possible equations of Y.

38. A Quadratic function given by Y = $(x - a)^2 + p$, the coordinates of the minimum point is given by (5, -10).

  a. Find the vales of a and p

  b. find the coordinates of X and Y intercepts of the function ,

  c. hence Sketch the function Y = $(x - a)^2 + p$

39. write an expression for the discriminant of the equation given by $ax^2 + bx + c = 0$

40. Show that the equation $x^2 + 4x + 1 = 0$ has two distinct roots.

41. Show that the equation $2x^2 + 4x + 7 = 0$ has no real roots

42. Show that the equation $x^2 - 2x + 1 = 0$ has equal roots.

43. Show that the equation $x^2 + bx - 1 = 0$ has two distinct roots for any value of b.

44. Show that the equation $x^2 + 2px + p^2 = 0$ has equal roots for any value of p.

45. Show that the equation $x^2 + 2p + 2p^2 = 0$ has no real roots for any value of p.

46. a. Show that the equation $x^2 + (2 + p)x + p = 0$ **can not** have **equal roots** for any value

  b. given p = -1 , solve the equation by giving the answers in surd form.

47. Equation $x^2 + (2+k)x + k + 1 = 0$ has equal roots,

   a. Find the values of K.

   b. Hence solve the equation

   c. Sketch the graph $y = x^2 + (2+k)x + k + 1$

48. Equation $px^2 + qx + p = 0$ has equal roots,

   a. Write an expression for the discriminant

   b. Hence write the value of **q** in terms of **p**. And solve the equation.

49. Given that the discriminant of the quadratic equation $(x+3)(x-\lambda) = 0$ is zero

   a. What you can predict about the roots of this equation.

   b. Find the values of λ

   c. Hence sketch $y = (x+3)(x-\lambda)$, Show the coordinates of X intercepts , Y intercepts and Minimum points.

50. Find the values of λ where the equation, $8(\lambda - 2x)(\lambda + x) = 25 + 16\lambda$, has equal roots. Hence solve the equation.

51. An equation given by $x^2 + 2x = \lambda - 3$

   a. write an expression for the discriminant in terms of λ

   b. Find the range of λ where the equation gives two real roots.

   c. Hence solve the equation for the minimum integer value of λ.

52. $\dfrac{\lambda}{x+1} = \dfrac{x}{x+2}$ has equal roots. Find the value of λ.

53. An equation given by $x^2 + bx + 3 = 0$, has two distinct real roots. find the range of b giving your answer in surd form.

54. An equation given by $2x^2 + bx + 6 = 0$, has two distinct real roots. find the range of b giving your answer in surd form. Hence solve The equation for the minimum integer value of b.

55. Find the value of λ where the equation, $8(\lambda - 2x)(\lambda + x) = 25 + 16\lambda$, has two distinct roots.

    a. write an inequality in terms of λ

    b. hence deduce the range of λ

    c. Solve the equation for the minimum positive integer value of λ.

55 The equation, $2px^2 + (1 + p)x + 1 = 0$, has two distinct roots.

    a. write an inequality in terms of p

    b. hence deduce the range of p.

    c. Solve the equation for the minimum positive integer value of p.

56. Equation, $t^2 + (2 + p)t = p$, has two distinct roots for t.

    a. show that $(p + 4)^2 > 12$

    b. Show that $a^2 - b = (a - \sqrt{b})(a + \sqrt{b})$

    c. Hence or otherwise solve the inequality $(p + 4)^2 > 12$, to find the values of p in surd form.

57. Equation, $(1 - p) x^2 + 2(1 + p)x + (1\text{-}p) = 0$ has two distinct real roots.

  a. Show that $(p + 1)^2 - (1 - p)^2 > 0$

  b. Hence show that $P > 0$

58. A tank of water has a height of 5 units , length is (L+3) units and width (4-L) units.

  a. write an expression for the volume (V) of the tank in terms of L

  b. find the maximum value of the volume

  c. sketch a graph of V against L.

59. A rectangle has a length of (p –x) and width of x . the area of the rectangle is A.

  a. Write an expression for A

  b. Given that the Maximum value of A is 25 , Find the value of P.

  c. Hence sketch a graph of A against x.

60. A rectangle has a length of (p –x) units and width of x units . the area of the rectangle is 25 square units.

  a. Write an expression for which relates X and p

  b. show that $p \geq 10$

61. Quadratic curve given by $Y = px^2 + qx + 3$. the gradient of the tangent to the curve at x = -1 is 3 . and the gradient of the tangent to the curve at x = 2 is 1.

a Sate $\frac{dy}{dx}$ in terms of p and q

b Find two equations relating p and q

c hence find the quadratic equation of the curve , find the minimum value of Y .

62. Quadratic curve given by $Y = px^2 + qx + 1$. the gradient of the tangent to the curve at x = 1 is -3 . and the gradient of the tangent to the curve at x = -4 is 1.

a Sate $\frac{dy}{dx}$ in terms of p and q

b Find two equations relating p and q

c hence find the quadratic equation of the curve , find the minimum or maximum value of Y .

63. Quadratic curve given by $Y = x^2 + qx + p$. the gradient of the tangent to the curve at x = 1 is 3 . And the coordinates of minimum point the curve is (- 0.5 ,3.75) .

a find $\frac{dy}{dx}$ and find the value of q

b find the possible values of p

c sketch graphs for the possible values of p

64. Quadratic curve given by $Y = px^2 - x + q$. the gradient of the tangent to the curve at x = 1 is 3 . And the minimum value of Y is -3 .

a find $\frac{dy}{dx}$ and find the value of p

b find the coordinates of the minimum point

c. Find the value of q.

d find the coordinates where the curve cuts x and Y axis.

65. The coordinates of the minimum point of a quadratic graph is given by (2 , 5). The graph cuts the Y axis at (0, 7). Find the equation of the quadratic graph.

66. The coordinates of the maximum point of a quadratic graph is given by (2 , 5). The graph cuts the Y axis at (0, 3). Find the equation of the quadratic graph.

67. A quadratic function given by Y = f(x) passes through (0,5) also given that when x = -1, $\frac{dy}{dx} = 0$. and x = 2 , $\frac{dy}{dx}$ = -3 ,

a. find the equation of the quadratic function.

b. solve Y = 0,

b. Sketch Y = f(x) , Clearly showing the intercepts , and the coordinates of minimum /maximum point.

68. A quadratic curve given by Y = x^2 + bx - 3 . the equation of the line of symmetry given by X = 1

a. By completing the square or otherwise find the value of b

b. find the minimum value of Y

c. Solve Y = 0

69. . A quadratic function given by Y = f(x) passes through (0,-2) also given that when x = 1, $\frac{dy}{dx} = 0$. and x = -2 , $\frac{dy}{dx}$ = 1 ,

a. find the equation of the quadratic function.

b. solve Y = 0,

b. Sketch Y = f(x)

70. A quadratic curve given by Y = x^2 + x - c . the minimum value of Y is equal to -3

a. find the equation of the line of symmetry

b. find the value of c.

c. Solve Y = 0 , leave the answers in surd form.

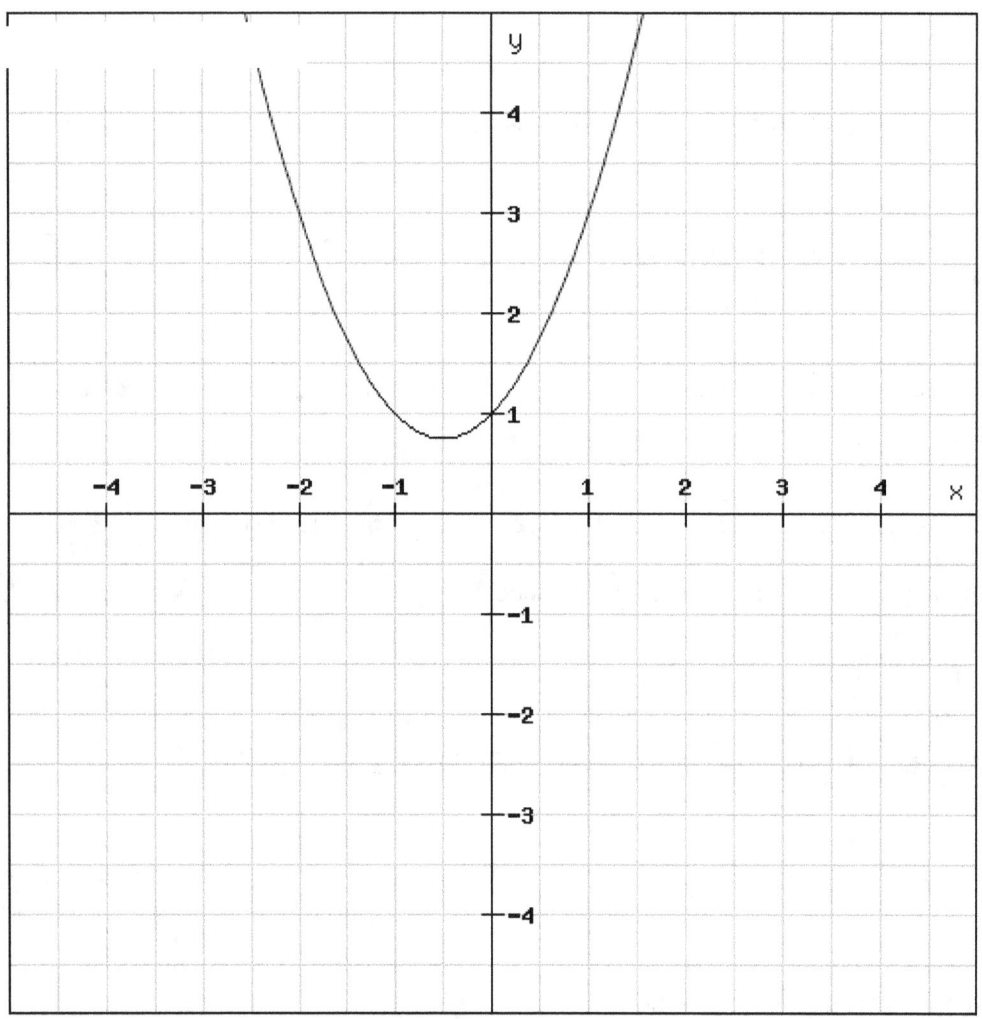

71. Above shows a quadratic graph **f(x)** drawn to real scale. Find the equation of the graph. And Sketch **2f(x) – 3** on the same grid.

72. A quadratic curve given by Y = ax^2 + x - 1 . the minimum value of Y is equal to -3 ,

a. show that $12a^2$ -4a - 1 = 0

b. find the possible values of a

c. Solve Y = 0 .

73. Given that , $\dfrac{x-k}{x-3} + \dfrac{2}{x-2} \equiv \dfrac{P(x+q)^2 +r}{(x+2)(x+1)}$

    a. find p, q and r in terms of k .

    b. Given that $\dfrac{x-k}{x-3} + \dfrac{2}{x-2} = 0$, has equal roots find the possible values of K

74. Given that $ax^2 + x + b$ is a complete square . Show that $4ab = 1$

Solve the following simultaneous equation.   **3A**

1. $2x + 5 = 2y$, $3x - 2y = 12$ **2.** $2x + 7y = 2$, $x + 2y = 2$

3 $-2x + 9 = 2y$, $-3x + 2y = 5$ **4.** $2 + 7y = 5x$, $x + 2y = -12$

5. $-3t = 2p$, $-3t + 2p = 12$ **6.** $-2p + 10 = 7q$, $-3p + 2q = 8$

7 . $\frac{x}{y} = -3$, $7x + 2y = 5$ **8.** $\frac{x-3}{y} = 5$, $\frac{x}{4y} = 3$,

9. $\frac{3}{p+6} = \frac{-2}{q+4}$, $\frac{3-q}{p-6} = 1$ **10.** $X - y = 2 = x^2 + y^2$ **11.** $X + y = 2 = x^2 + y^2$

12. $X + 2y = 0 = x + 2y^2$ **13.** $5 = x + 2y^2$, $y - x = 1$ **14.** $17 = 2x^2 - y^2$, $x^2 + y^2 = 10$

15. $xy + x^2 = 0$, $y - x = 2$ **16.** $(x + y)^2 + x = 1$, $x^2 + 2xy + y^2 - 2x = 7$

17. Lines $X + 3y = 11$ and $Y = 3X + 4$ intercept at points P , Find the coordinates of P .

18. $y^2 - 2xy = 8$ and $y = 3x - 1$ intercepts at p and Q. find the coordinates of P and q.

19. $Y = \lambda x$ and $Y = x^2 - \lambda x + 3$ intercepts at x = 3 . find the value of λ. Hence find the coordinates of the points of interception.

20. $Y = \lambda^2 x$ and $Y = x^2 + \lambda x + 4$ intercepts at x = 2 . find the values of λ in surd form. Given that $\lambda > 0$. Find λ and the coordinates of interception.

21. $Y = x + 3\lambda$ and $Y = \lambda + 1 - 2x$ intercepts at y= 2 . find the value of λ. Hence find the coordinates of the points of interception. Find λ and the coordinates of interception

22. $Y = 3\lambda + x$ and $Y = -6 - 2\lambda x$ intercepts at y = 2 . find the values of λ. Hence find the coordinates of the points of interception. Given that $\lambda > 0$. Find λ and the coordinates of interception

23. $Y = \lambda x$ and $Y = x^2 - \lambda x + 3$ intercepts at point p and q. Given that $\lambda > 0$

    a. find the range of λ

    b. Find the points P and q for the minimum integer value of λ

24. $Y = \lambda x$ touches the curve $Y = x^2 - \lambda x + 1$ at point p, where $\lambda > 0$

    a. find the value of λ

    b. hence find the coordinates of p.

25. given that, The circle $x^2 + y^2 = 4$ and the line $Y = \lambda x + 1$ intercepts at p and q.

    a. Show that Show that $(1 + \lambda^2) x^2 + 2x\lambda - 3 = 0$

b. by considering the discriminant, show that that λ can take any value.

c. Find the coordinates of P and Q when $\lambda = -1$

Solve the following inequalities (show the sollutions on a number line) ,

26. $X + 3 > -9$ **27.** $-2x + 1 > 4$ **28.** $4 - 4x \geq 2x$ **29.** $6 + 3x < -x$ **30.** $-y - 3 < -2y$

31. $(x + 3)(x-3) \geq 0$ **32.** $-(x+1)(3x -6) > 0$ **33.** $x^2 + 5x + 6 > 0$ **34.** $x^2 + 5x < 0$ **35.** $x^2 - 36 > 0$

36. $81 - x^2 < 0$ **37.** $x^2 + 2x > 6$ **38.** $5^x > (25)^{-x+4}$ **39.** $5^{x^2} \geq (125)^{-x}$

40. $3^{x^2} < (27)^{-x+4}$

41. find the range of X which satisfies both $X + 5 \geq 3x - 3$ and $5x \geq x - 8$

42. find the range of X which satisfies both $-X + 4 \geq -2x - 2$ and $x < 1$

43. find the range of y which satisfies both $3y - 4 \geq y$ and $-y + 3 < -3y$

44. find the range of X which satisfies both $x^2 - 2x - 3 < 0$ and $x < 1$

45. find the range of X which satisfies both $x^2 - 2x - 3 < 0$ and $-x^2 + 2x > 0$

46. find the range of y which satisfies both $2y^2 - y - 3 > 0$ and $-y^2 + 2y > 0$

47. Sketch $(x-1)(X-2)$, and $(2-X)(x-5)$ on the same grid and find the X values which satisfies both

$(x-1)(X-2) \geq 0$ and $(2-x)(x-5) \leq 0$

48. Sketch (x-1)(X-2) , and (X-2)(x-5) on the same grid and find the X values which satisfies both (x-1)(X-2) ≥ 0 and (x-2)(x-5) ≥ 0

49. Sketch (x-2)(X-5) , and (X-3)(x-4) on the same grid and find the X values which satisfies both

(x-1)(X-2) ≤ 0 and (x-2)(x-5) ≤0

50. Find the values of X which satisfies -(x-2)(X-5)≥ 0 and   3 < X<4

51. Find the ranges of X and Y which satisfies both expressions given by, y = x – 3 and   y + 2x > 0

52. Find the ranges of X and Y which satisfies both expressions given by, x + 2y = -4 and y – 3 > -x

53. Find the ranges of X and Y which satisfies both expressions given by, $5^x > (25)^{x-y}$ and $3^y = (27)^{1+x}$

54. Find the ranges of X and Y which satisfies both equations given by $2^{x^2+y^2} < 16^{\frac{-x}{4}}$ and

Y = X +1

55 . A quadratic expression is given by λx^2 (3+ λ)x + 3 , where λ is a positive constant

    a. Factorise the expression

    b. hence or otherwise solve the inequality λx^2 (3+ λ)x + 3 > 0 in terms of λ

56. A quadratic expression is given by x^2 +2 λx + λ^2 , where λ is a positive constant

    a.   Factorise the expression

b. hence or otherwise show that $x^2 + 2\lambda x + \lambda^2 \geq 0$ for all the values of λ.

57. By a suitable substitution, **a**. Solve the inequality, $-Z^2(Z^2 - 1) \geq -2$ for z^2 **b**. Hence Solve the inequality for **Z**.

58. Solve the inequality $(Z^2 - 4)(Z^2 + 1) < 0$

59. The equation $x^2 + \lambda x + 1 > 0$ has no real roots . Find The range of λ

60. The equation $\lambda x^2 + \lambda x + 1 > 0$ has two real roots . Find The range of λ

61. Solve the inequality which satisfies, $(x - 1)^2 < 4x + 1$ and $-(x - 4)^2 > 2 - x$

62. The equation $\lambda x^2 + x + \beta = 0$ has two distinct roots.

　a. Write an inequality relates λ and β

　b. Also given that $\lambda - \beta = 0$. show that $1 - 4\beta^2 > 0$, hence Find the ranges of λ and β

63. The equation $x^2 + \lambda x + \beta = 0$ has equal roots.

　a. Write an equality relates λ and β

　b. Also given that $\lambda + \beta > 0$. Hence find the ranges of β and λ.

64. The inequality $x^2 +(\lambda - 3)x > 0$ has two solutions given by $X < 0$ and $X > 4$. Find the value of λ.

65. The expression given by, $Y = \lambda x^2 +(\lambda - 3)x -3$

    a. Factorise the expression in terms of λ

    b. hence solve the inequality $Y > 0$ in terms of λ where $\lambda > 0$

    c. Given when $Y > 0$, $X < -1$ and $X > 1$, Find the value of λ

66. The length of a rectangle given by $(x-3)$, and the width is given by $x-5$.

    a. If the area has to be greater than 24 , Form an inequality in terms of x.

    b. Solve it and find the range of X.

67. The length of a rectangle given by x , and the width is given by $5-x$.

    a. If the area has to be greater than 6 , Form an inequality in terms of x.

    b. Solve it and find the range of X

68. The length of a rectangle given by $(x-3)$, and the width is given by $x-5$.

    a. If the area has to be greater than 24 , Form an inequality in terms of x.

    b. and If the Perimeter has to be less than 24 , Form another inequality in terms of x.

    c. Solve both of the inequalities and find the range of X which satisfies both conditions.

69. Length of a Rectangular area is $(2x+3)$ and the width is given by $(x-4)$. The Area of the rectangle has to be greater than 20 and the perimeter has to be less than 22. Find the range of X , which would satisfy both these conditions.

70. Find the ranges of X and Y , which will satisfy $XY \geq 10$, and $Y - X = 9$

4: Sketching Curves and Transformations.

Sketch the following curves showing the X and Y intercepts (coordinates) : 4A

1. $Y = X(X+3)$ **2.** $Y = x(X-1)(x+4)$ **3.** $Y = x^2(X+1)$ **4.** $Y = x(X-1)^2$

5. $Y = (x-1)(X+1)(X-3$ **6.** $Y = (2x+4)(X-2)$ **7.** $Y = x(5x-3)(2x+7)$

8. $Y = -(x+4)(x-5)$ **9.** $Y = -x^2$ **10.** $Y = -x^2(X-1)$ **11.** $Y = 3x(X-1)^2$

12 . $Y = -2x(X-1)(x+4)$ **13.** $Y = -4x^2$ **14.** $Y = -x^2 - x$

15. $Y = x^3 + 5x^2 + 6x$ **16.** $Y = -2x^3 + 4x^2 + 6x$ **17.** $Y = x^2 - 1$

18. $Y = 4x^2 - 25$ **19.** $Y = x^3 - x$ **20.** $Y = 81x - x^3$

21. $Y = \dfrac{1}{x}$ **22.** $Y = \dfrac{-1}{x}$ **23.** $Y = x^{-1}$ **24.** $Y = -x^{-1}$ **25.** $Y = \dfrac{5}{x}$ **26.** $Y = \dfrac{-5}{x}$

27. $Y = x^3$ **28.** $Y = -x^3$ **29.** $Y = -2x^3$

30.   Sketch $Y = \dfrac{1}{x}$ and $Y = x^2(x+1)$ on a same grid . Hence show that $\dfrac{1}{x} = x^2(x+1)$ has two real roots. Where $x \neq 0$.

31. Sketch $Y = \frac{1}{x}$, and $Y = -x(x+1)$ on a same grid . Hence show that $1 + x^2(x+1) = 0$ has one real root. Where $x \neq 0$.

32. Sketch $Y = -x^{-1}$, and $Y = x^2(x+1)$ on separate grids . given that $x^3(x+1) + 1 = 0$ has no real roots . sketch the graphs on a same grid

33. a Sketch $f(x) = -x^{-1}$, and $g(x) = x^2(x-2)$ on a same grid .

 b. hence deduce the number of possible roots which satisfies $x^3(x-2) + 1 = 0$ in accordance with your sketch. Where $x \neq 0$

 c. Find f(1) , g(1) , f(1.5) and g(1.5)

 e. Hence the deduce the exact number of roots for $x^3(x-2) + 1 = 0$ between $0 \leq x \leq 2$

34. $f(x) = x$, and $g(x) = x^2(x+1)$

 a. use Algebra to find the roots of $f(x) = g(x)$

 b. Hence Sketch f(x) and g(x) clearly showing the coordinates of the roots of f(x) = g(x).

35. a Sketch $f(x) = x$, and $g(x) = (x-2)^2(x+1)$ on a same grid .

 b. use algebra to show that $f(x) - g(x) = x^3 - 3x^2 - x + 4$

 c. Hence deduce the number of possible roots which satisfies $x^3 - 3x^2 - x + 4 = 0$ according to your sketches.

36. a Sketch $f(x) = x^3$ and $g(x) = (x-2)^2(x+1)$ on a same grid .

 b. Hence comment on the possible number of real roots of f(x) – g(x) = 0

37. a Sketch $f(x) = -2x^3$ and $g(x) = (x)^2(x-1)$ on a same grid .

 b. solve the equation $3x^3 - x^2 = 0$

 c. Hence find the coordinates where f(x) and g(x) intercepts.

38.a. Sketch $f(x) = (x-1)^2(x+1)^2$ and $g(x) = x^2 - 1$ on a same grid

 b. using algebra show that $(x-1)^2(x+1)^2 = (x^2-1)^2$

 c. Hence or otherwise solve $(x-1)^2(x+1)^2 = x^2 - 1$

 d. Hence find the coordinates where f(x) and g(x) intercepts.

39. a. Sketch $f(x) = x^2(x-1)^2(x+1)^2$ and $g(x) = 4x^2$ on a same grid

 b. using algebra show that $x^2(x-1)^2(x+1)^2 = x^2(x^2-1)^2$

 c. Hence or otherwise solve $(x-1)^2(x+1)^2 = 4(x)^2$

 d. Hence mark the coordinates where f(x) and g(x) intercepts

40. a. Factorise $(x^3-x)^2 - 4x^2$

 b. Hence solve , $(x^3-x)^2 - 4x^2 = 0$

 c. sketch $f(x) = 4x^2$, and $g(x) = (x^3-x)^2$, showing and stating the coordinates where f(x) = g(x).

41. a. Factorise $4x^2 - 16$

 b. sketch $f(x) = 4x^2 - 16$ and $g(x) = x^2$ on a same grid.

c. find the coordinates where f(x) = g(x) , keep the answer in surd form.

42. Two functions given by   f(x) = $0.5x^{-1}$, and g(x) = $8x^2(x-1)^2(x+1)^2$

a. show that f (0.5) < g (0.5)

b. Hence or otherwise sketch f(x) and g(x) clearly showing the interceptions between f(x) and g(x)

c. deduce the exact number of real roots $16x^3(x-1)^2(x+1)^2$ -1 = 0   where x > 0

43. Two functions given by   f(x) = f(x) = $\frac{5}{x}$ and g(x) = x^2 -4x + 6.5

a. By completing the square or otherwise find the minimum value of g(x).

b. write the coordinates of the minimum point of g(x)

c. find g(0)

d. show that f(2) = g(2)

e. using the answers to a,b,c and d Sketch f(x) and g(x) on a same gird

f. hence predict the number of real solutions for the equation $2x^3$ -8x^2 + 13x -10 = 0

44. Two functions given by   f(x) = $10x^2-1$, and g(x) = $x^2(x-2)^2(x+2)^2$

a. Write the coordinates where f(x) and g(x) would cut x and Y axis.

b. Write the coordinates of the minimum point of f(x)

c. show that f(1) = f(-1) = g(1) = g(-1)

d. using the answers to parts **a,b**, and **c** ,Sketch f(x) and g(x) on the same grid.

45. Two functions given by $f(x) = -x^3$, and $g(x) = (x-2)^2 - 8$

   a. find the coordinates of the minimum point of g(x)

   b. find the coordinates where g(x) cut the x and y axis.

   c. find f(2) and g(2)

   d. by using the answers to a , b and c   Sketch f(x) and g(x) on a same gird clearly showing their points of interceptions.

46. Solve X(X+5) + 6 = 0 , hence sketch Y = $-6x^{-1}$ and y = x+ 5 clearly showing and stating the coordinates where they intercept each other.

47. Given that, $f(x) = x^2$, find :

   a. f(x-3)   **b.** f(2x)   **c .** 2f(x)   **d.** f(-3x)   **e.** 2f(0.5x)

   f. -2f(-3x)   **g.** f(-3x) - 4   **h .** 3- f(-x)   **I.** -3f(x-1) -3   **J .** f(5x -3)   **K .** f(x^2)

   L. f $(2x^2$ +3)   **M.** - f ($\frac{x}{3}$ + 1)   **N.** -f($3x^2 - 2$) + 3   **O.** $(f(3x-1))^2$ **P.** f($\frac{2x-3}{3}$)   **Q.** $\frac{f(x-1)}{3}$

   R. f(x) + 2f(x-3)   **S.** $\frac{f(x) + 2f(x-3)}{4}$ **T.** $\frac{f(x) + 2f(x-3)}{4}$ - 3

47. Given that, $f(x) = 3x^2 - 27$, Sketch f(x) stating the coordinates of the minimum point . hence sketch the following curves

   a. f(x-3)   **b.** f(2x)   **c .** 2f(x) - 3   **d.** f(-3x)   **e.** 2f(-0.5x)

   f. -2f(-3x)   **g.** f(-3x) - 4   **h .** 3 - f(-0.25x)   **I.** -3f(x-1) -3   **J .** f(5x -3)   **K** f($\frac{2x-3}{3}$) ,

   L. $\frac{f(x-1)}{3}$ -3 **M** 1 + 2 f(-x)

48 Given that f(x) = 2x -3 and g(x) = x^2 , Sketch the following

a. f(x) + g(x   **b**. f(x) – g(x)   **c**. f(x-1) + f(x)   **d**. g(x) - g(x-1) **e**. f(x) – 2g(x)

f. f(x-3) + 3   **g**. 2f(x+3) -5   **h**. g(x-4) – 4   **I**. g(2x-2) **K**. –f(x)   **L**. -2g(x)

49. The point P(**a,b**) lies on the curve given by f(x). find the new coordinates of the point P in terms of **a** and **b** after the following transformations.

a. f(x-3)   **b**. f(2x)   **c** . 2f(x) - 3   **d**. f(-3x)   **e**. 2f(-3x)

f. -2f(-3x)   **g**. f(0.5x) - 4   **h** . 3 - f(-0.25x) **I**. -3f(x-1) -3 **J** . f(5x -3) **K** f($\frac{2x-3}{3}$) ,

L. $\frac{f(x-1)}{3}$ -3 **M** 1 + 2 f(-x) **o**. f(x) + 2

50. The point P(**-3,4**) lies on the curve given by f(x). find the new coordinates of the point P after the following transformations.

a. f(x-3)   **b**. f(2x)   **c** . 2f(x) - 3   **d**. f(-3x)   **e**. 1 - 2f(-3x)

f. -2f(-3x)   **g**. f(-3x) - 4   **h** . 3 - f(-x) **I**. -3f(x-1) -3 **J** . f(5x -3) **K** f($\frac{2x-3}{3}$) ,

L. $\frac{f(x-1)}{3}$ -3 **M** 1 + 2 f(-x) **o**. f(x) + 2

51. Explain graphically what do you mean by a **Stretch** and a **translation** of a curve along a given axis.

52. The curve f(X) undergoes the following translations and/or stretches . find the resulting curve in terms of f(X).
    a. translated along the X axis by +3 units
    b. translated along the X axis by -3units
    c. stretched along Y axis by a scale factor of 3

d. stretched along X axis by a scale factor of 3

e. stretched along Y axis by a scale factor of 1/3

f. translated along the X axis by +3 units followed by a stretch along Y axis by a scale factor of 3

g. translated along the Y axis by +3 units followed by a stretch along X axis by a scale factor of 3

h. translated along the X axis by +3 units followed by a stretch along Y axis by a scale factor of 3

I. Stretched along the X axis by scale factor of 3 followed by a translation along X axis by + 3units

J. Stretched along the Y axis by a scale factor of 3 followed by a translation along Y axis by +3 units

K. Stretched along the X axis by scale factor of 1/3 followed by a translation along X axis by - 3units

L. Stretched along the Y axis by scale factor of 1/3 followed by a translation along X axis by - 3units

53. Given that f(x) $= (x - 1)^2$ and g(x) $= 4x^2$

    a. Show that f(x) $= \frac{1}{4}$g(x-1)

    b. solve $(x - 1)^2 = 4x^2$

    c. Hence sketch f(x) and g(x) on a same grid.

54. Given that f(x) $= (x + 1)^2$ and g(x) $= 4(4x + 9)^2$

    a. Find f(x) - g(x), hence sketch f(x) - g(x)

    b. use part a and Sketch f(x-1) - g(x-1)

    C. Use Algebra to find the solutions for the equation, f(x-1) - g(x-1) = 0

55, f(x) $= - 4(Kx-1)(X+4)$, where K is a constant. Given g(x) $= 4f(\frac{x}{2}) \equiv -x^2 -4x + 16,$

    a . find the value of K

    b . Hence sketch f(x) and g(x)

    c. Sketch g$(\frac{x}{2})$,

    d. Find the maximum value of g$(\frac{x}{2})$,

56, A cubic curve C cuts the X axis at points P(3,0), Q(-1,0) and R(-3,0) . also given the Curve cuts Y axis at (0,-7) . Sketch the curve C and find the equation of the curve C in the form of Y $= ax^3 + bx^2 + cx + d$

57, A cubic Curve cuts the x axis at P(2,0) , Q(-1,0) . there's a double root at point p. and the Curve cuts Y axis at (0,-3). . Sketch the curve C and find the equation of the curve C in the form of $Y = ax^3 + bx^2 + cx + d$

57, A cubic Curve Y = f(x) cuts the x axis at P(2,0) , Q(-1,0) . there's a double root at point p. when x > 2 , $\frac{dy}{dx} > 0$. Sketch the curve f(x)

58, A cubic Curve Y = f(x) cuts the x axis at P(-3,0) , Q(1,0) . there's a double root at point p. when x < -3 , $\frac{dy}{dx} > 0$. Sketch the curve f(x)

59, A Curve Y = F(x) = -2x(x-3)(x+1) is transformed into 2f(x+1) - 3 via three consecutive transformations. State the three transformations . Hence sketch 2f(x+1) - 3

60, A Curve Y = F(x) = $3x^2$(x+1) is transformed into -3f(x-1) -1 via three consecutive transformations. State the three transformations . Hence sketch -3f(x-1) -1

61, A Curve Y = F(x) = x^2(x+1) is transformed into f(2x-3) -1 via three consecutive transformations. State the three transformations . Hence sketch f(2x-3) -1

62, A Curve Y = F(x) = -2x(x-3)(x+1) is transformed into 2f(3x+1) via three consecutive transformations. State the three transformations . Hence sketch 2f(3x+1)

63, Curve C given by Y = f(x) , after a Stretched along the X axis by scale factor of 1/3 followed by a translation along X axis by - 3units the curve f(x) transformed into 4(x-3)(x+1) . find f(x)

64, Curve C given by Y = f(x) , after a translated along the X axis by -2 units. followed by a stretched along x axis by a factor of 2, the curve f(x) transformed into -3(2x-4)(x+1) . find f(x)

65, Curve C given by Y = f(x) , after a Stretched along the Y axis by scale factor of 1/3 followed by a translation along X axis by - 1unit the curve f(x) transformed into x^2(x+2)(5x+1) . find f(x)

66, Curve C given by Y = f(x) , after a Stretched along the y axis by scale factor of 5 followed by a translation along y axis by +3units the curve f(x) transformed into x^2(3x-3)(x-1) . find f(x)

67, Curve C given by Y = f(x) , after a Stretched along the y axis by scale factor of 3 followed by a translation along x axis by +3units the curve f(x) transformed into 3x(x+2)(x-5) . find f(x)

68, Curve C given by Y = f(x) , after a translated along the X axis by -1 unit. followed by a stretched along x axis by a factor of 0.5, the curve f(x) transformed into (2x-4)(x+1) . find f(x)

69, Curve C given by Y = f(x) , after a translated along the X axis by -3 units. followed by a stretched along x axis by a factor of 2, the curve f(x) transformed into (x-2) (5x-4)(x+1) . find f(x)

70, A cubic Curve Y = f(x) cuts the x axis at P(-3,0) , Q(1,0) . there's a double root at point p. when x = -3 , $\frac{dy}{dx}$ = -1 . find f(x) in the form of K(x-a)(x-b)(x-c) . Sketch f(x)

71, A cubic Curve Y = f(x) cuts the x axis at P(-3,0) , Q(-1,0) . there's a double root at point Q. when x = 2 , $\frac{dy}{dx}$ = 3 . Sketch the curve f(x)

72, A cubic Curve Y = f(x) cuts the x axis at (0,0) , (p ,0), and (q,0) . And also given the coefficient of the x^3 is 1.

    a.   Show that f(x) = $x^3 -(p+q)x^2$ + pqx

    b.   Find $\frac{dy}{dx}$

    c.   Given that $\frac{dy}{dx}$ = 1 where x = 0 , find a relationship between **p** and **q**

    d.   Also given that that $\frac{dy}{dx}$ = 0 where x = 1 , find another relationship between **p** and **q**

    e.   Hence find **p** and **q** and factorise f(x) completely

    f.   Sketch f(x)

72, A cubic Curve Y = f(x) cuts the x axis at (0,0) , (p ,0), and (2p,0) . where p > 0 And also given the coefficient of the x^3 is 1.

    a. Show that f(x) = $x^3 -(3p)x^2 + 2p^2$x

    b. Find $\frac{dy}{dx}$

    c. Given that $\frac{dy}{dx}$ = 11 where x = 1 , Show that $2p^2$ -6p -8 = 0 and sketch f(x)

73, A cubic Curve Y = f(x) cuts the X axis at (p ,0), and (1-p,0) and goes through the origin , And also given the coefficient of the x^3 is 1.

    a. Show that f(x) = $x^3 -x^2 +(p - p^2)$x

    b. Find $\frac{dy}{dx}$

    c. Given that $\frac{dy}{dx}$ = 3 where x = -1 , find the possible values of p

    d hence Sketch f(x)

74, A cubic Curve $Y = f(x)$ cuts the X axis at $(p, 0)$, and $(1,0)$ and goes through the origin ,

    a. Show that $f(x)$ can be written in the form of $K(x^3 - (p+1)x^2 + px)$, where K is a constant

    b. Find $\frac{dy}{dx}$

    c. Given that $\frac{dy}{dx} = 3$ where x = -1 , and when x = 1 , $\frac{d^2y}{dx} = 0$ find the possible values of p and k.

    d hence Sketch $f(x)$

75. Given that $f(x) = 16(2x-1)^2$ and $g(x) = 4x^2$

    a. factorise $f(x) - g(x)$

    b. hence or otherwise sketch $f(x) - g(X)$ and $f(x+1) - g(X+1)$

    c. solve $f(x) - g(x) = 0$

76. Sketch $y = (x-2)^3$ and $y = -(x+2)^3$, Hence solve $(x-2)^3 + (x+2)^3 = 0$

77. Show that $a^3 + b^3 = (a+b)(a^2 - ab + b^2)$, Hence Factorise $(x+1)^3 + (1-x)^3$ and sketch $Y = (x+1)^3 + (1-x)^3$.

78. Show that $a^3 + b^3 = (a+b)(a^2 - ab + b^2)$, Hence Factorise $(2x+1)^3 + (1-x)^3$ and solve , $(2x+1)^3 + (1-x)^3 = 0$

79. Show that $a^3 - b^3 = (a-b)(a^2 + ab + b^2)$, Hence Factorise $x^3 - (x-1)^3$ and show that there aren't any real solutions for $x^3 - (x-1)^3 = 0$.

80. a. Show that $a^3 - b^3 = (a-b)(a^2 + ab + b^2)$,

b. show $x^3 - (ax-1)^3 = \{x(1\text{-}a)+1\}\{(a^2+a+1)x^2+(2a+1)x+1\}$

c. Show the discriminant of $(a^2+a+1)x^2+(2a+1)x+1=0$ is lesser than negative

d. Hence show that $x^3 - (ax-1)^3 = 0$ has **only one real solution** for **any value of** a

81 . Given that $a^3 - b^3 = (a\text{-}b)(a^2+ab+b^2)$

    a.   Show that $x^3 - (x-a)^3 = a(3x^2-3ax+a^2)$
    b.   Show that the discriminant of $3x^2-3ax+a^2$ is greater than zero
    c.   Hence Show that the equation $x^3 - (x-a)^3 = 0$ has two real roots for any value of a
    d.   Given that a = 1 , sketch the graph $Y = x^3 - (x-1)^3$

82. The curve C given by f(x) = 3(x -1)(2 –X)(X +1) transformed to a curve given by g(x) = -3x(3-x)(x+2) by two consecutive transformations . and the g(x) transformed into h(x) = 6x(x-3)(x+2) - 4

    a.   State the two transformations that would transform from f(x) to g(x)
    b.   State the two transformations that would transform from g(x) to h(x)
    c.   Sketch f(x) and h(x)

83. A cubic function f(x) is given below . **a.** find the equation of the cubic function . **b.** sketch 2f(x+3) - 4 on the same grid. **d.** state the number of solutions to the equation

    f(x)- 2f(x+3) +4 = 0

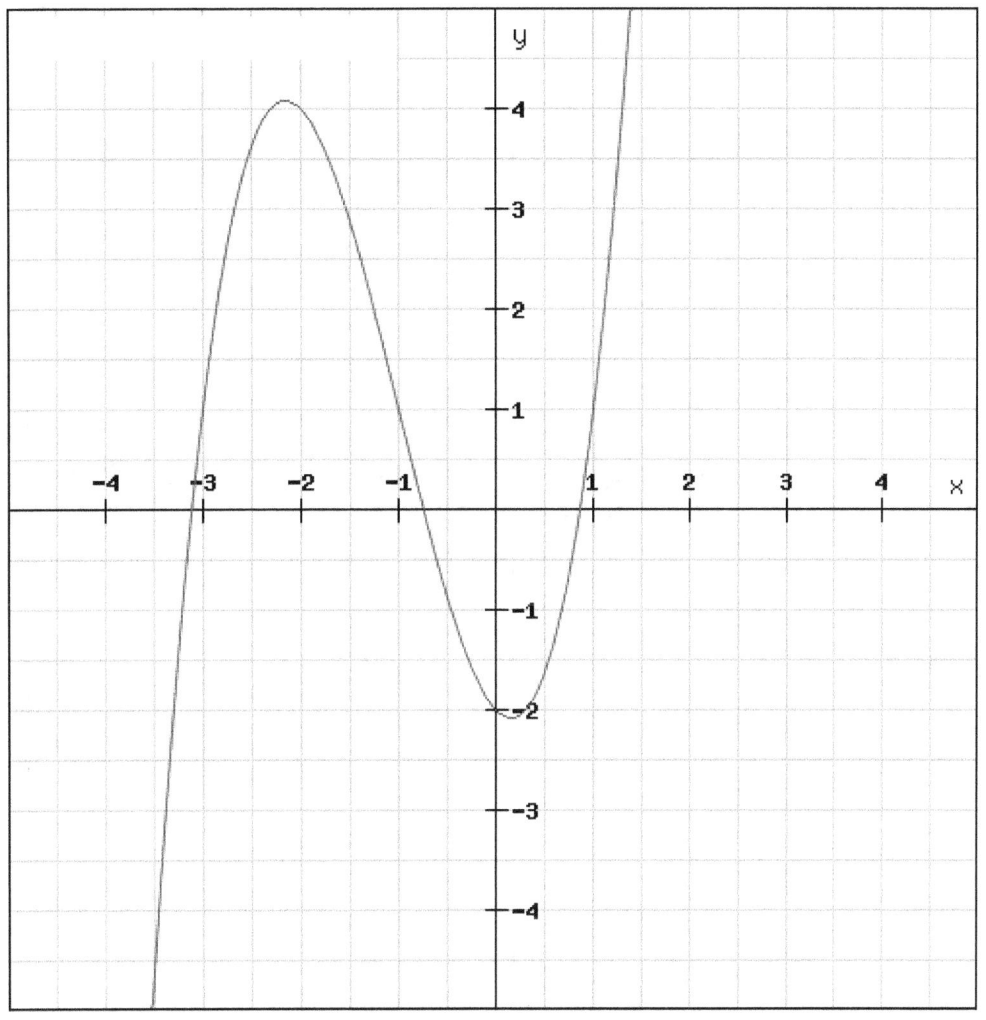

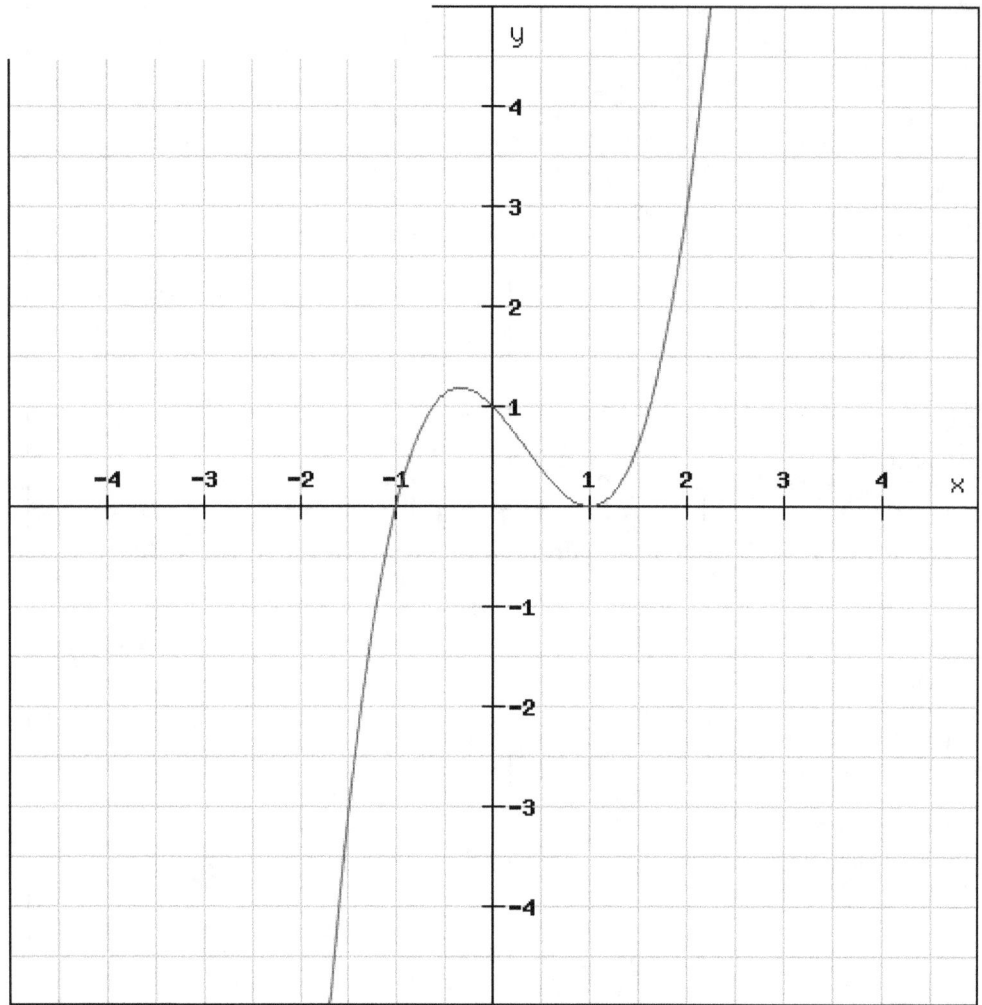

84. The curve **C** shows above given by $f(x) = 3x^2(X + 1)$ transformed to a curve given by $g(x) = -3x(x-1)^2$ by two consecutive transformations . And the g(x) transformed into $h(x) = 6x(x+1)^2$ by further consecutive transformations

   a. State the two transformations that would transform from f(x) to g(x)

   b. State the two transformations that would transform from g(x) to h(x)

85, The curve C given by f(x) = (x -1)(2 –X)(X +1) transformed to a curve given by g(x) = -2x (X+1)(2X +3) by two consecutive transformations . And the g(x) transformed into

h(x) = 2x(-x+3)(-2x+3) +2 by further consecutive transformations

a. State the two transformations that would transform from f(x) to g(x)

b. State the two transformations that would transform from g(x) to h(x)

86. A cubic function f(x) is given below . **a.** find the equation of the cubic function .

b. sketch f(-x+3) - 4 on the same grid.   **d.** state the number of solutions to the equation f(x)- f(-x+3) +4 = 0

5: COORDINATE GEOMETRY - STRAIGHT LINE

5A

1. Consider the Following equations of straight lines.

(i) . Find gradients of the of the lines

(ii) . Find the X and Y intercepts

(iii). Hence sketch the lines .

a. $Y = -3x + 5$ **b.** $2Y = 6 + 4X$ **c .** $X - 3 = Y$ **d.** $Y + 1 = -3x$ **e.** $Y = -2(X-3)$
F. $0 = 3(2y - 4x +1)$ **g.** $-1 = 2(Y -3x+1)$

2. Find the distance and the midpoint between the a pairs of coordinate points given below
   a. (1,3) and (6,5)   **b.** (-4,0) and (5,7)   **c.** (-9 ,7) and (0,0)   **d.** (-6,5) and (4,5)
   e. (a-3, 4) and (7+a, -3)   **f.** (-5,-6) and (-6,-7)   **g.** (-6 , 0) and (-1,-2)   **h.** (p-5, -5) and (5-p , 7)

3. Line L1 passes through points A (1,1) and B (-1 , 5) .
   (i)   find the length AB .
   (ii)   Find the gradient of the line.
   (iii)   Find the equation of the line L1
   (iV) Line L2 is parallel to the line L1 and crosses the X axis at point P (-3 ,0) and Y axis at
point Q find the equation of the line and the area of the triangle POQ. Where **o** is the origin.

4. Line L1 passes through points A (3,1) and B (-1 , 5) and crosses Y axis at pint S and X axis at R
     (I). Find the length AB .
     (II). Find the gradient of the line.
     (III). Find the equation of the line L1
     (IV). Line L2 is parallel to the line L1 and crosses the Y axis at point P (7,0) and X axis
     at point Q. find the equation of the line L2 and Find the Area of the trapezium PQRS

5. The Line L1 Passes the X axis at A $(0,a)$ and Y at B$(b,0)$. given that $a + b = 9$ and the Area of the triangle AOB is 4. Where O is the origin. Find the values of a and b. hence the deduce equations of L1. Also show that the coordinate $(-3, 10)$ cannot be on the line L1.

6. The Line L1 Passes the X axis at A $(0,p)$ and Y at B$(q,0)$. given that $p + q = 9$ and the Area of the triangle AOB is 4. Where O is the origin.

   (i). Find the values of p and q.
   (II) Hence show that there are two possible equations for L1, and find the two equations in the form of $Ax + By + C = 0$
   (iii) Given that the two equations of L1 crosses at point N, find the coordinate of point N.

7. The Line L1 Passes the Y axis at A $(0,3)$ and X at B$(4,0)$. and Line L2 passes the Y axis at point P $(0,2)$ and X at Q$(7,0)$ and they intersect at point N.
   (i). Find the equations of the lines L1 and L2 in the form of $Ax + By + C = 0$
   (II) Sketch the lines on a X Y
   (iii) Find the coordinates of the point N, leave the answers as fractions
   (iv) Find the Area **BOPN**, where O is the origin, give the answer as a fraction.

8. A Straight line given by $aY = aX - 1$, the line crosses the Y axis at A $(0,5)$ and X axis at point B
   (i). Find the gradient and the value of a. **(ii)** find the X intercept of the line. **(iii)** Find the coordinates of the midpoint of between A and B. **(iv)** calculate the exact distance AB

9. A Straight line given by $CY = C(X-1) - 5$, the line crosses the Y axis at A$(0,-2)$ and Y at point B
   (i). Find the gradient and the value of C. **(ii)** find the X intercept of the line. **(iii)** Find the coordinates of the midpoint of between A and B. **(iv)** calculate the distance AB
   (V) Calculate the shortest distance between the **origin** and the **line AB**

10. A Straight line given by Y = aX + a-1 , the line crosses the Y axis at P (0,5) and X axis at Q.

    (i)   Find the gradient and the value of a.

    (ii)   Hence solve aX +a − 1 = 0

    (iii)   Show the length PQ $= 5\sqrt{\dfrac{37}{6}}$

    (iv)   The mid-point of PQ given by R, find the area of the triangle given by PRO , where O refers to the origin.

11. Straight lines Y = AX +B and Y = BX − A   intercept at point P (1,2).

    a.   Write two equations which links A and B.

    b.   Hence find the values of A and B.

12. Straight Line L , makes an angle of Θ (Anti clockwise) with the X axis. And the Y intercept of the line is (0,-3). Given that tanΘ $= \dfrac{\sqrt{2}}{3}$.

    a.   find the equation of the line L. hence find the exact value of the Y coordinate where x = -2

    b.   Find the exact coordinates where the line crosses the X axis.

13. Straight lines Y = AX +B and Y $= \dfrac{X}{A} - B$ intercept at P (1,1).

    a. Write two equations which links A and B.

    b. Show that $(A - 1)^2 = 0$ Hence find the values of A and B.

    c.   Find the distance between P and the origin.

14. Straight lines L1 and L2, Y = AX +B and Y $= \dfrac{X}{A} - B$ intercept at (1,1).

    a .Write two equations which links A and B.

    b .Show that $(A - 1)^2 = 0$

    c .Hence find the values of A and B.

d. Find the vales of X where $AX + B = 0$

15. Straight lines $Y = AX + B$ and $Y = \dfrac{X}{A} - B$ intercept at $(2,3)$.

   a. Write two equations which links A and B.

   b. Show that $A^2 - 3A + 1 = 0$

   c. Hence find the exact values of A and B.

16. Straight Line L, makes an angle of Θ (anti clockwise) with the X axis. And passes through the point $(\sqrt{7}, 2)$. Given that $\tan\Theta = \dfrac{5}{\sqrt{7}}$. find the equation of the line L in the form of $Ax + by + c = 0$. hence find the exact value of the X coordinate where $Y = -2$

17. Straight lines $Y = AX + B$ and $AY = X - AB$ intercept at $(p,1)$. Where A and B are real variables.

   a .Write two equations which links A and B in terms of p.

   b .Show that $PA^2 - 2A + P = 0$

   c. Deduce that $1 - p^2 \geq 0$

   d. Hence write the range of possible values for P.

   e. Given $p = 0.5$, find the exact values of A and B.

18. $Y = 2x - 1$ and $2Y - 3X + 1 = 0$ intercept at (a,b). Show that $a = 1$ and $b = 1$.

19. Show that Straight lines $Y = AX + B$ and $AY = -X + AB$ will not intercept at $(1,1)$

20. $Px + qY + r = 0$ intercepts X axis at A $(-3,0)$ and the line $Lx + My + N = 0$ at point B $(3,0)$. and

they intercept each other at C (2, 7). Where P , Q , r , L , M and N are constants.

    a. Make sketches of $Px + qY + r = 0$ and $Lx + My + N = 0$ with relevant coordinates.

    b. Find the values of P , Q , r , L , M and N

    c. Find the area of the Triangle ABC.

21. $Y = aX + b$ intercepts X axis at A (P,0) and the line $Y = cX + d$ cuts the X axis at point B (5,0) . and they intercept each other at C (2, 7). The area of the triangle ABC is 49 square units.

    a .Make a Sketch of the lines $Y = ax+b$ and $Y = cX +d$ with relevant coordinates

    b. Find the possible values of P

    c. hence find the values of a,b,c and d.

22. Line L1 create an angle of Θ (Anti clockwise) with the X axis and Line L2 create an angle of β (Anticlockwise) with the X axis. And Line L1 and L2 meet where (2,3). Given that $\tan Θ = -2$, and $\tan β = 0.5$ Find the equations of L1 and L2.

23. Line L1 , given by $Y = 2x + p$ passes through X axis at point A . line L2 passes through the X axis at point B . Lines L1 and L2 meet at point C (3,5) . given that AC = BC.

    a. Find the value of p

    b. State the value of the gradient of line L2.

    c. Find the equation of the line L2

    d. Workout the coordinates of point A and point B.

    e. Find the Area of the triangle ABC

    f. Another Line, L3 passes through point C and its perpendicular to the X axis , state the equation of line L3.

24. Line L1 , given by $3Y = -2x + p$ passes through Y axis at point A . line L2 passes through the Y axis at point B . Lines L1 and L2 meet at point C (3,5) . given that AC = BC.

   a. Find the value of p

   b. State the value of the gradient of line L2.

   c. Find the equation of the line L2

   d. Workout the coordinates of point A and point B.

   e. Find the Area of the triangle ABC

   f. Another Line, L3 passes through point C and its perpendicular to the Y axis , state the equation of line L3.

25 . Show that all three lines $y = 2x - 3$, $-2y + 5x + 1 = 0$ and $Y = -x + 5$ will not intercept each other at a one point.

26. Given that line L1, $y = 2x + a$,Line L2 $Y = -2x + b$ and line L3, $X = C$ lines intercept at point P(3,5). find the values of a,b and C. Given also that L1 crosses Y axis at point A and Line L2 at point B. Workout the Area of the triangle APB.

27 .Line L1 : $aY = bX + 1$ and Line L2 : $Y = -2x + 2$ are two parallel lines . L1 crosses the Y axis at point A and L2 at point at point B. The length between A and B is 5 units. Find the possible values of the constants a and b.

28. Line L1 : $Y = 2x + 5$ crosses the X axis at point A , and the L2 $Y = -2X + 8$ crosses the X axis at point B. L1 and L2 meet at point M .

     a. Draw a sketch of lines L1 , L2 and L3 clearly showing the points A,B and M
     b. State the coordinates of M
     c. Line L3 : $Y = K$ where $K < 6.5$ crosses the Lines L1 and L2 at P and Q respectively . The area of the trapezium ABPQ 6 square units. State the coordinates of the points P and Q in terms of K

d.   show that the length of PQ $= \frac{13-k}{2}$

e.   Find the value of K

29. Line L1 passes through the points A (2,1) and B (-4,5).

    a.   Find the coordinates of the mid point (C)

    b.   Find the equation on the line L1 in the form of px + qy + r = 0

    c.   Line L2 passes through the midpoint and crosses the Y axis at D(0, P), where P >0, length of CD is $\sqrt{10}$. Find the value of P and hence find the equation of the line L2.

30.   Points A , B and C are given by (2,3) , (4,5) and (P,Q) . Given that ABC is a right angled triangle where AB   is the hypotenuse

    a.   Find the length of the hypotenuse

    b.   Find the length of AC and BC in terms of P and Q.

    c.   Hence show that , $P^2 + Q^2$ - 6P - 8Q + 23 = 0

    d.   Given that Q = 3 find the values of P.

31.   Lines L1 and L2 are perpendicular to each other. And line L1 given by Y = -3x +4 and line L2 passes the   Point (-1 1).

    a.   Find the equation of the line L2 in the form of ay + bx + c = 0 .

    b.   Show that point A (2,2) lies on the line L2

    c.   Find the coordinates where L1 and L2 intersect.

    d.   Hence find the shortest distance between point A and the Line L1.

32. Lines L1 given by Y = X and Line L2: Y = mX + 7   are perpendicular to each other .

    a. Find the value of m

    b. lines L1 and L2 intercept at P . find the coordinates of P

    c. Given that triangle PQR has an area of 50 square units , where Q is the Y intercept of the line L2 and R is a point on the line L1. Find the length of PR in surd form .

33. line L1 : Y = Ax + B and Line L2 : Y = BX − A lines are perpendicular to each other at p. Line L1 goes through   point (2,1) .

   a. find the values of the constants A and B.

   b. Hence justify that there are **two** possible pairs of straight lines

   c. find the   coordinates of p.

34. Lines L1 and L2 are parallel to each other. And the shortest distance between the lines is $\sqrt{5}$ units . Given that   L1 : Y = 2x -5 and L2   crosses Y axis at (0,K)

   a. Show that k^2 + 10k = 0

   b. find the values of K

   c. Hence find the possible equations of the line L2

35. Lines L1 : Y = -4x -4 and line L2 : Y = mx + c are perpendicular to each other . Line L1 crosses the Y axis at A and Line L2 Crosses the X axis at B (K,0) . the length AB is 5 units.

   a. Find the value of m

   b. find the possible values of k

   c. Hence or otherwise find the values for C.

36. Line L1 crosses the X axis at point A (p,0) and Y axis at point B (0,q) . line L2 : 3x + 5y -1 = 0 Bisects the length perpendicularly AB at   point R.

   a.   Form two equations which relates P and Q.
   b.   Hence find the values of p and q.
   c.   Find the equation of the line L1
   d.   Find the length AB

37. Line L crosses the X axis at point A (X,0) and Y axis at point B (Y,0) . given that X + Y = D and the length AB is 10 .

a. show that $2X^2 - 2DX + D^2 - 100 = 0$

b. by considering the discriminant of the equation from part (a) Show that $100 \geq D^2$

c. Hence state the range of D

d. Given that D = 0 find the values of X and Y.

e. Hence find the possible equations of the Line L.

38. The straight line given by L 1: $5x - 2y + 1 = 0$.

a. show that point A (1,3) lies on L1

b. point B(P,Q) also lies on line L1 and given that AB = 10 , find the exact values of P and Q

39. Find the shortest distance between the line Y = 3x -2 and the point (7,-1)

40. Three points A(X+1, x) , B(2X +1 , x-1) and C(3x+1, 3) are collinear. (a) Find the value of X. (b) Show that C is the mid point of AB. (c) Find the equation of the line AB in the form of Ax +by + c = 0

41. Points P(A,B) and Q(C,D) are lies on the Line L1: $Y - 3x + 4 = 0$. The length PQ = 10. Given that A > 0

a. Show that $(A - C)^2 = 10$

b. given that C = 1 , find the value of A in surd form.

c. Hence find the value of B and D

42. Lines L1 : Y = 5x − 3 and L2 : 0= 3x − 2y + 1 intersect at the point A . the point B(P,Q) lies on L1 and point C (R,S) Lies On L2. BC and AC perpendicular to each other.

a. Find the gradient of the line BC

b. Find the coordinates of the point A

c. Show that 13R -34P + 21 = 0 ,

43. Two lines L1 : $y = 3x - 3$ and L2 : $Y = x + 5$ intercepts at point P. line L1 crosses X axis at Q and L2 crosses the X axis at R.

   a. find the equation of the line which passes through P and perpendicular to X axis.

   b. find the area of the triangle PQR

   c. equation L3 passes through point P L1 crosses the X axis at S. given that the area of PQS is 9. Find the possible equations of L3 in the form of $AX + By + c = 0$

44. Points A , B and C are given by (2,3) , (4,5) and (P,Q) . Given that ABC is a right angled triangle where AB is the hypotenuse

   a. Find the length of the hypotenuse
   b. Find the length of AC and BC in terms of P and Q.
   c. Hence show that , $P^2 - 6P + 23 = - Q^2 + 8Q$
   d. By Completing the square of the right hand side of the above equation Show that
   $P^2 - 6P + 7 \leq 0$

45. Lines L1 and L2 are parallel to each other. And the shortest distance between the lines is 1 unit . Given that L1 : $Y = 2x - 5$,

   a. Show that $k^2 + 10k + 20 = 0$

   b. find the values of K in surd form

   c. Hence find the possible equations of the line L2

6: Differentiation

Basics

1. Sketch $Y = \frac{12}{x} - 1$. the points A(2,p) , B(4,q) , C(6,r) , D (12,s) lies on the curve.
   a. Find the values of p , q , r and s
   b. Find the gradients of the lines AD, BD, and CD
   c. Comment on the answer to part b
   d. Find the derived function of Y , hence find the value of the gradient of the tangent at point D
   e. Find $\frac{dy}{dx}$ where x = -12
   f. Hence find the equation of the line of the tangent where x = -12 in the form of Ax + by + c = 0

2. Given that $f(t) = t^2 - 3t - 4$
   a. Find $f'(t)$
   b. Hence find the value of the gradient of f(t) at t = -1
   c. Workout the equation of the tangent to the curve where t = -1

3. Points A(1, p) , B(1.1, q) and C (1.2, r) lies on the curve given by , $Y = Kx^2$, The gradient of the chord AB is equal to 21.
   a. Find the values of p ,q and r in terms of K
   b. Hence find the value of K
   c. Find the gradient of the chord AC
   d. Comment on the value of the gradient at point A in comparison to the gradients of the chords AB and AC
   e. Find exact value of the gradient at A by differentiation.

4. Given that f(x) $= -6x^2$ - 3x - 5

    a. Find the first and the second derivative of f(X)

    b. Hence find $\frac{dy}{dx}$ when x = -1

    c. Workout the equation of the normal to the curve when x = -3

5. Given that f(t) $= 6t^2$ - 3t $-\frac{2}{t}$ + 1 and g(t) $= \frac{2}{t}$

    a. Find $\frac{df(t)}{dt}$ and $\frac{d^2f(t)}{dt}$

    b. Find the equation of the normal to f(t) when t = 1

    c. Find the first derivative of f(t) + g(t)

    d. Hence find the value of t where the first derivative is zero for f(t) + g(t)

6. **Differentiate with respect to x**

    a. 2x   **b.** −x   **c.** 0   **d.** -5   **e.** -3x   **f.** $-x^2$ **g.** $6x^3$ **h.** $\frac{1}{5x}$ **I.** $\frac{x^{-4}}{4x}$

    J. x(x-1)   **K.** x(2x-1)(X+1)   **L.** $x^{-3}(x^2 +1)$ **M.** 5(x- $\sqrt{x}$)   **N.** -4(2x + $\sqrt[3]{x}$)

    O. $x^{-2}(x - 4x)$ **P.** $2(\frac{x}{x+1})^{-1}$ **q.** $(\frac{x}{x+1})^{-2}$ **r.** $x(\frac{2x}{2x+1})^{-1}$ **S.** (x+1) $(\frac{x}{x+1})^{-1}$

    T. $\frac{x^2+2x+1}{X+1}$,   **U.** $\frac{x^2-1}{x+1}$ **V.** $\frac{3x^2-3}{x+1}$ **X.** $\frac{x^4-1}{x^2+1}$ **Y.** 2 (x+1) $(\frac{x}{x+1})^{-1}$

    Z. $3x^{\frac{1}{5}}$ + $4x^3$ - 4x + 3

7. **Differentiate with respect to t,   K is a constant**

    a. $\frac{1}{Kt} + \frac{2}{t^2} + 3t^{-3}$ **b.** $\frac{\sqrt{5kt}}{t}$ **c.** $\frac{K+\sqrt{3t}}{5t}$ **d.** $\frac{(t-K)(t-3)}{t^3}$ **e.** $\frac{(t-1)(t-\sqrt{t})}{5t}$

    f. $\frac{t^{\frac{2}{3}}}{kt}$ **g.** $4t^{\frac{1}{3}}(t^{\frac{2}{3}} - Kt^{\frac{-1}{3}})$ **h.** $K(t-K)(t-5)(\frac{t}{2}-4)$ **l.** $\frac{(t-k)(t-\sqrt{k})}{t^{-1}}$

8. A curve **C** is given by f(X) = $x^2 - 4x + 1$, point **p** and **Q** lies on the curve

    a. Find the first and the second derivatives.

    b. Find the coordinates where the value of the first derivative is zero.

    c. Find the equation of the tangent to the curve at point **P** (2, -3)

    The tangent to the curve at point **Q** passes the point   **R**(2,-6) . Given that the X coordinate of the point **Q** is **M** ,

    d. workout $\frac{df(x)}{dx}$ at point Q in terms of M.

    e. State the Y coordinate of point **Q** in terms of **M**

    f. Show that the tangent at point Q can be written as Y = (2M-4)x + 1 - M^2

    g. show that , M^2 - 4M + 1 = 0

9. A curve **C** is given by Y = $\frac{1}{3}x^3 + 3x^2$ + 5x +1 , tangents to the curve   at points P and Q has zero gradients.

    a. Find an expression for the gradient function $\frac{dy}{dx}$

    b. find the coordinates of the points P and Q

    c. Hence deduce the equations of the lines of the tangents at P and Q.

    d. State the equations of the normal to the curve at points P and Q

    e. Find the exact values of **x** coordinates where the gradient of C will be equal to one.

10. The gradient function of the curve **C** is given by $f'(x) = \frac{1}{x} - x + 3$, $X \neq 0$

    a. There are two points A and B on the curve C the tangent to the gradient equal to -3 . find the **X** coordinates of the points A and B

    b. Given that the y coordinate of A is 1 and the Y coordinate of B is 3 , find the equations of the normal at point A and B.

    c. The tangent at point A meets the normal at point B at point C, and the tangent at point B meet the normal at point A at point D. with or without a aid of a sketch state a possible name for the quadrilateral ABCD

    d. Find the coordinates of C and D

    e. Find the length of the diagonal AB

11. Derived function of the curve C is given by $f'(x) = \frac{1}{x^3} - 10x - 3$, $X \neq 0$

    a. Find the second derivative of C

    b. Given that the normal to the curve at x = 2 , passes through the origin. Find the equation of the normal in the form of Ax + By + C = 0

12. The straight line L , which passes through the point A (-2,3) cuts the curve C perpendicularly at point B(M,N). The curve C given by $f(X) = x^2 + x + 1$.

    a. Write the full coordinates of B in terms of M

    b. Write an expression for the gradient of the line L in terms of M

    c. show that $2M^2 - M = 0$

    d. Deduce the possible values of M

    e. Given that M ≠ 0. Find the equation of the line L.

13. Two curves given as $f(x) = x^2$ and $g(x) = (x - k)^2$, where K > 0 , is a constant.

    a. Find the coordinates of point P where the curves intercept in terms of K

    b. Find the gradients of the curves at point P in terms of K

    c. Given that the Curves cut perpendicularly each other at P, find the value of K

14. Given , $Y = x^3 - 2ax^2 - 7x + b$, where a and b are constants.
    a. Write an expression for the gradient at any given x
    b. Find the gradient at x = 0
    c. The curve passes through the point R(-1,3) , and the gradient at R is equals to 0.5 .
       Find the values of a and b

15. Given that f(Θ) = Θ(Θ-a)(Θ-b), a and b are constants.
    a. Find $f'(Θ)$ in terms of a and b
    b. Given that $f'(Θ) = \frac{7}{4}$ where Θ = 0 , show that ab = $\frac{7}{4}$
    c. F(Θ) satisfies the coordinate (-2,1) , Find the values of a and b as fractions
    d. Hence sketch f(Θ)

16. Given that f(x) = (x-2)(x-3)(X+1), Find the equations of the tangent and the normal at X = 2

17. Volume of a balloon at "t" time given by the relationship , V(t) = $3t^{\frac{2}{3}}$ - 8t + 1
    a. find $\frac{dv}{dt}$
    b. find the rate of change in the volume of the object when t = 8
    c. find the time when $\frac{dv}{dt} = 0$, comment on meaning of this answer
    d. find the range of t where the balloon is expanding

18. f(Θ) = $-2Θ^{\frac{5}{2}}$ + $Θ^2$ + 3 ,
    a. find $f'(Θ)$
    b. given that $f'(p) = f'(Q) = 0$, find the values of p and Q.

19. Curve C given by , f(x) $= \frac{3(x-1)^2}{x}$,

a. Find f '(x)
b. Find the range of X , where f '(x) > 0
c. find the values of X where the curve C's rate of change against x is zero.

20. Area Of an object given by $A = 2Cr^2 - \frac{4\pi r^2}{10}$, where C is a constant

a. Find $\frac{dA}{dr}$
b. Given that when r = 0.5 , $\frac{dA}{dr} = 2$, find the value of C in terms of π

21. f(x) $= \frac{K}{x^4} + \frac{4}{x^2} - c$

a. Given f($\sqrt{3}$) $= 1$, show that $9C - K = 3$
b. Also given that f'(1) = -1 , find the values of K and C
c. Find the equation of the normal to the curve where x = 1

22. The gradient function of the curve C is given as , f '(x) $= \frac{Kx}{1-3x}$, where K is a constant.
Also Given that f '(k) $= \frac{1}{4}$ find the values of k.

23. The gradient function of the curve C is given as , f '(θ) $= \frac{K\theta-4}{1+3\theta^2}$, where K is a constant.
Also Given that f '(k) = -2 find the values of k.

24. A Quadratic curve C given as Y $= (x - 3)^2 + 7$, points P and Q satisfies the curve C. the tangents at points P and Q meet at the origin.

a. Differentiate Y
b. Find the possible equations of the line of the tangents

25. Sketch Y $= -x^2 - 4$, The tangents to the curves at points P and Q meets at the origin.

65

    a.   Workout the equations of the tangents .

    b.   Find the coordinates of the points P and Q

    c.   Find the area of POQ triangle.

26.   Curve C given by $f(x) = -4x^3 - Kx^2 - 12x$, K is a constant , K > 0

    a.   Given that $f'(p) = 0$, show that $Ap^2 + Kp + C = 0$, where A and C has to be found

    b.   Also given that **that there's only one possible value for p** , find the value of K

    c.   Hence find the value of p

    d.   Find the value of f(p)

    e.   deduce the equation of the tangent and the normal to the curve at point P

27. Curve C given by $f(\theta) = \dfrac{\theta^3 - K\theta^2 + 10}{\theta}$, K is a constant ,

    a.   Differentiate f(θ) with respect to θ and find f''(θ)

    b.   f'(2) = f''(-2) find the value of K

28. Differentiate with respect to x, $f(X) = x^{\frac{1}{3}} + 4x^{\frac{1}{2}}$ - 6x , find f'(64) and f'(8) in its simplest forms

29. Differentiate with respect to X , $f(x) = x^{\frac{1}{3}}(Px^{-\frac{2}{3}} + 4x^{\frac{1}{3}} - 6px^{\frac{-1}{3}})$, where P is a constant find $f'(p^3)$, leave the answer in terms of P in the simplest form.

30. $f'(1) = 2f'(2)$, where $f(x) = x^2 + BX + 1$, find the value of B. hence find the value of $f'(1)$

31. $f'(1) = -2f'(2)$ and $f'(3) = 1$ where $f(x) = Ax^2 + BX + 1$, find the values of A and B. hence find the value of $f'(1)$

32. $f(\theta) = \dfrac{3\theta^2 - \theta}{2\theta}$, find f'(θ) and f''(θ)

33. $f(x) = x^2 + 5X + 1$, find f'(x-1) and f''(x-1)

34. $f(x) = x^{\frac{1}{3}}(x^{\frac{1}{3}} - 1)(x^{\frac{1}{3}} + K)$, given that $f'(8) = f''(-8)$. find the value of K . and find the equation of the normal to f(x) where x = 8 in the form of y = mx +c

35. A curve C give by $f(x) = x^2 + 5x + 6$.

    (a) find the range of x values where the curve C has a positive gradient .

    (b) Find the equations of tangent and the normal at point P(1,12) in the form of Y = mx + c

    (c) The tangent and the normal cut the X axis at R and S respectively. Find the area of the triangle PQR

36. A curve C given by $f(x) = x^2 + 5x + 6$.

    (a) find the coordinates of the points P and Q ,where f(x) = 2.

    (b) Hence find the equations of the tangents at points P and q. the tangents meet at point R.

    (c) find the coordinates of the point R and then find the area of the triangle PQR.

37. A parabola given by $y = (K-4)x^2$. where K is a constant k > 0

    a. Find $\frac{dy}{dx}$

    b. Given that at x = k , $\frac{dy}{dx}$ is 10 . find the value of K

    c. Hence find the equation of the normal to the parabola at X = K

    d. The normal to the parabola at x = k cuts the parabola to the second time at x = p , show that p satisfies the relationship given by $250p^2 + 25P - 510 = 0$

38. **Find the first and the second derivative (f '(t)and f ''(t))with respect to t**

    a. $\frac{1}{t} + \frac{2}{t^2} + 3t^{-3}$ **b.** $\frac{\sqrt{5t}}{t}$ **c.** $\frac{x+\sqrt{3t}}{5t}$ **d.** $\frac{(t-1)(t-3)}{t^3}$ **e.** $\frac{-(t-1)(t-\sqrt{t})}{5t}$ **f.** $\frac{t^{\frac{2}{3}}}{t}$

    c. $4t^{\frac{1}{3}}(t^{\frac{2}{3}} - t^{\frac{-1}{3}})$ **h.** $(t^{\frac{-2}{3}} - 1)(t^{\frac{-2}{3}} + 1)t$ **l.** $2(t^{\frac{-2}{3}} - 1)(t^{\frac{-2}{3}} + 1)t^{\frac{4}{3}}$

39. A curve is given as , $Y = x^3 + 5x^2 + 7x + 4$

    a. Find the coordinates where $\frac{dy}{dx} = 0$

    b. Find $\frac{d^2y}{dx}$

    c. Hence find the equations to the normals where $\frac{dy}{dx} = 0$

    d. Find the shortest distance between the two normals

40. A curve is given as , $Y = x^3 + 5x^2 + 7x + 4$

    a. Find the coordinates where $\frac{dy}{dx} = 0$, hence deduce the number of stationary points

    b. *Find $\frac{d^2y}{dx}$, hence deduce the nature of the stationary points

    c. find the equations to the tangents where $\frac{dy}{dx} = 0$

    d. Find the shortest distance between those tangents

41. The derived function of curve C is given as $f'(x) = x^{\frac{2}{3}}(x - 8)$

    a. Find the X coordinates of the possible stationary points

    b. *at x = 8 there's a stationary point classify the stationary point by taking $f''(x)$

    c. Given the curve passes through the point (-1,5) , find the equation of the tangent when x = -1 in the form of y = mx + c

42. The derived function , $f'(x)$ of the curve C is given as $f'(x) = 3x^{\frac{1}{3}}(x^{\frac{1}{3}} - 2)(x^{\frac{1}{3}} + 2)$

    a. Find $f'(8)$

    b. Find the equation of the normal to the curve where x = 8

    c. Solve $f'(x) = 0$, and deduce all the X coordinates of the stationary points of C

43. $Y = x^3 + 5x^2 + 7x - 1$, find the range of values where $\frac{dy}{dx} > 0$

44. $Y = x^3 - 9x$, find the range of X where $\frac{dy}{dx} < 0$

45. $Y = x^3 - 9x^2 + 1$, find the range of values of X where Y is decreasing against x

46. $Y = 4x^3 + x^7$, show that Y never decreases against X.

47. $Y = 4x^{-3} + x^{-7}$, $x \neq 0$, Show that for all X values $\frac{dy}{dx} < 0$

48. $Y = Kx^2 - 3X$, given that $\frac{dy}{dx} > 0$ where X = -1, find the range of K

49. $Y = K^2x^2 + 6Kx - 1$, given that $\frac{dy}{dx} \leq 0$ where X = -2, find the range of K hence find $\frac{dy}{dx}$ at x = -2 for the smallest positive integer value of K

50. Curve C given by, $Y = Kx^3 + 5x^2 + 5X + 4$, K is a constant

    a. Find $\frac{dy}{dx}$ and $\frac{d^2x}{dx}$

    b. Given that there's only one point of X where $\frac{dy}{dx} = 0$, show that $K = \frac{5}{3}$

    c. Hence find the coordinates where $\frac{dy}{dx} = 0$

51. Curve C given by, $Y = Kx^3 + kx^2 + 5X + 4$, K is a constant. given that that there are two stationary points, Show that $k^2 - 15K > 0$. hence find the range of K and illustrate your answer on a number line.

7: INTEGRATION

1. Integrate the following expressions with respect to X

    a. 2x   **b.** $1 - 3X$ **C.** $-x$ **d.** -1   **e.** 0   **f.** $4x^2 - 1$ **g.** $-3x^3$

    h. $x^{-2} - 3x^{-3} + 3x$ **I.** $x^{\frac{2}{5}} + 2x^{\frac{2}{3}} + 1$ **J.** $x^0 + 2$ **K.** $\frac{x^{n+1}}{n-2}$,

    L. $3x^{-0.5} + x^{\frac{1}{3}} - 1$ **M.** $\frac{x+1}{x-1}$ **N.** $\frac{X+x^2}{x+1}$ **o.** 4 (x-1)(X+2)   **P.** $\frac{x^2+5x+6}{x+3}$, X ≠ -3

    Q. $2x^{\frac{1}{4}}(x^{\frac{1}{2}} - 3x^{\frac{3}{4}})$ **R.** $\sqrt{5X} - 5X$ **S.** $\frac{x^2+3}{\sqrt{x}}$ **T.** $\frac{3-\sqrt{x}}{x^2}$

2. The gradient of the curve f(x) given by , f'(x) = 2(x-1)(X+1)(x+2) ,
    a. find $\int f'(x)dx$
    b. Given that the curve C passes through the point (1,2) , find f(x)
    c. Hence find f($\sqrt{2}$) in surd form.

3. The gradient function of a curve f(x) is given by f'(x) = $4x^{\frac{2}{3}} - x^{\frac{1}{3}} + 2$, the curve passes through the point
    (1,2) . find f(x) and f(-8)

4. Given that $\frac{dy}{dx} = 2x - x^{\frac{1}{2}}$, find $\int \frac{dy}{dx}$ dx

5. Gradient of the curve Y = f(x) given by , $\frac{dy}{dx} = 2x^{-\frac{1}{3}} - x^{\frac{1}{2}}$, the curve passes through (8, 2)
    find the equation of the curve.

70

6. Find $\int \sqrt{x}(2x^{\frac{3}{2}} + x)\, dx$

7. Gradient of the cure $Y = f(x)$ given by, $\frac{dy}{dx} = Kx^{-\frac{1}{3}} - x^{\frac{1}{2}}$, k is a constant. The curve passes through $(0, 2)$ and $(1,4)$ find the value of K and the equation of the curve.

8. Gradient of the cure $Y = f(x)$ given by, $\frac{dy}{dx} = \frac{Kx^{\frac{1}{3}} - x}{x^{\frac{2}{3}}}$, K is a constant. The curve passes through $(-8, 2)$ and $(0,4)$ find the equation of the curve.

9. Curve $f(x)$ has a gradient function given by $f'(x) = 4x^{\frac{3}{4}} - Kx^{\frac{1}{2}} + x - 1$, given also $f'(16) = 2$

   a. Find the value of K

   b. $\int f'(x)\, dx$

   c. Given that the curve passes through the point $(1,5)$, find $f(x)$

10. Find $4\int(1 - t)(1 + 3t)\, dt$

11. Find $\int -5(3 - tx)(x - 3t)\, dt$

12. Find $\int (x^{\frac{1}{2}} + xt^{\frac{3}{2}} - 1)\, dt$

13. Integrate with respect to t, K is a constant

   $a. \int 7 \left(\frac{1}{K} + \frac{2}{t^2} + 3t^{-3}\right) dt$ **b.** $\int \frac{\sqrt{5kt}}{t} dt$ **c.** $\int \frac{K+\sqrt{3t}}{5t} dt$ **d.** $\int \frac{(t-K)(t-3)}{t^4} dt$

   e. $\frac{(t-1)(t-\sqrt{t})}{5t}$ **f.** $\int 4t^{\frac{1}{3}} (t^{\frac{2}{3}} - Kt^{\frac{-1}{3}}) dt$ **g.** $\int K(t - K)(t - 5)(\frac{t}{2} - 4) dt$

h. $\int \frac{(t-k)(t-\sqrt{k})}{t^{-1}} \, dt$

14. If $\frac{x^2}{4} + \frac{y^2}{9} = 25$, find $\int y^2 dx$ and $\int x^2 dy$

15. *if* $y^2 - x^2 - 2x - 1 = 0$, find $\int y^2 dx$ and $\int y dx$

16. *if* $-4x^2 + y^2 - 6y + 9 = 0$ find $\int -4x^2 dy$ and $\int y dx$

17. Factorise $3x^2 + 10x + 7$, hence find $\int \frac{3x^2 + 10x + 7}{2x+2} dx$

18. $\int \frac{-3x^2 + 4x + 7}{3x-7} dx$

19. The Curve C given by f(Θ) . the point P is on the curve , the gradient function of the curve given by

$f'(\Theta) = \frac{(\Theta-2)(\Theta-K)}{\Theta-1}$,

a. given that f '(1) = -1 , find the value of K and $\int f'(\Theta) d\Theta$

b. Given the curve C satisfies (2,3) find the equation of the curve C.

20. $f(x) = \frac{4x+4}{-2x^2+x+3}$

a. Show that $\frac{1}{f(x)} = \frac{3-2x}{4}$

b. Find $\int \frac{3-2x}{4} \, dx$

21. $g(x) = \frac{3x+2}{9x^2-4}$,

a. Show that $\frac{1}{g(x)} = 3 + 2x$

b. Find $\int \frac{1}{g(x)} \, dx$

22. The **gradient** of the **normal** of a function f(t) at **point p** given by g(t) $= \dfrac{2t^4}{(3t-1)(1-t)}$, t

    a. Write f '(t) at point P
    b. Find $\int f'(t)dt$
    c. F(t) satisfies (-2,0) , find f(t)

23. The **gradient** of the **normal** of a function f(t) **at point p** given by g(t) $= \dfrac{2t+3}{2t^2+5t+3}$

    a Find f '(t) at point P
    bFind $\int f'(t)dt$
    c.F(t) satisfies (-2,0) , find f(t)

24. F(x) $= 3x^{-2} - x^{\frac{1}{4}}$

    a. Find $\int$ F(x) dx
    b. Given that g(x) = $\int$ F(x) dx, Evaluate g(1) - g(0)

25. F(k) $= -2k^{\frac{1}{4}} - k$

    a. Find $\int$ F(k)dk

    b. Given that g(k) = $\int$ F(k) dk, Evaluate g($2^{\frac{1}{4}}$) - g(1) in surd form

26. f(x) = 2x + 3 , g (x) = $2x^2 + 5x + 3$, find **a.** $\int \dfrac{g(x)}{f(x)}dx$ **b.** $\int (f(x) + g(x))dx$
    c. $\int \dfrac{g(x)}{x^4}dx$ **d.** $-2\int f(x)g(x)dx$

4 : Arithmetic Series

1. Show that 4 , -2, -8 , -14 ..... is an arithmetic series.

2. Show that 10 , 11 , 14 , 15 .... can not be an arithmetic series.

3. Show that t +1, 2t -2 , 3t - 5 , 4t -8 ,..is an arithmetic series.   Given that   common difference is 4 , find the value of t . find the sum of first 5 terms.

4. The first term of an arithmetic series is -3 and the common difference is -3 . find the fifth term of the series.

5. The nth term of a Series given by U_n . Given that U_n = 2n-1.
   a. Find U_{n+1}
   b. Find U_{n+1} - U_n
   c. Hence deduce that the series must be an arithmetic
   d. Write the first 5 terms
   e. Find the sum of the first 20 terms

6. n th term of an arithmetic series is given by U_n = An + B, the 3rd term is -3 and 10 the term is 20.

   a. Find A and B
   b. Hence find U_1 and U_2
   c. Find S_{10}
   d. Find the Sum from the 3rd term to the 10th term

7. The nth term of a Series given by U_n . Given that U_n = -2n +5.

    a. Find U_{n+3}

    b. Find U_{n+4}

    c. Find U_{n+4} - U_{n+3}

    d. Hence write the first 5 terms

    e. Find the sum of the first 20 terms

8. There are 21 terms in an arithmetic series . The sum of the series is zero. Given that the first term is a, and common difference is d. and also given the 5th term is 6.

    a. Show that a = 10 and d = -1

    b. Write an expression for the n th term

    c. Hence Show that $\frac{U_{n+1}}{U_n} = \frac{10-n}{11-n}$

    d. Also show that $\frac{S_n}{S_{n+1}} = (\frac{n}{n+1})(\frac{10+U_n}{10+U_{n+1}})$

9. nth term of a series given by U_n , for a particular series , U_5 - U_6 = 3 and $\frac{U_5}{U_6}$ = 2 .

show that this series can't be an arithmetic series .

10. Arithmetic series given by U_n = -2n +5. Find $\sum_1^{10} U_n$ and $\sum_2^{12} U_n$

11. Arithmetic series given by U_n = **P**n +5. Given also that $\sum_3^{12} U_n$ = -21

    a. Find U_3 and U_{12} in terms of **p**

    b. Find $\sum_3^{12} U_n$ in terms of **p**

    c. Find the value of p

    d. Hence find U_{100}

12. Arithmetic series given by U_n = 2n - 5. Given also that $\sum_3^k U_n$ = 144

    a. Find U_3

    b. Find $\sum_3^k U_n$ in terms of k

    c. Find the value of k

13. Arithmetic series given by U_n = -10n - 5. Given also that $\sum_3^k U_n$ = -45k

    a. Find U_3

    b. Find $\sum_3^k U_n$ in terms of k

    c. Show that k^2 -7k -8 =0, and find the value of k

    d. Hence find the value of $\sum_3^8 U_n$

14. Arithmetic series given by U_n = 3n - 15. Given also that $\sum_4^k U_n$ = 15k

    a. Find U_4

    b. Find $\sum_4^k U_n$ in terms of k

    c. Show that $3k^2$ -57k +54 = 0, and find the value of k

    d. Hence find the value of $\sum_4^{18} U_n$

15. First term of an **arithmetic** series given by **a** , the common difference is given by **d** . and the number of terms given by **n** . the nth term given by U_n .

    a. State an expression for U_n in terms of **a, d** and **n**

    b. Given for a particular **arithmetic** series $\sum_5^{10} U_n$ = 0 and $\sum_8^{14} U_n$ = 70 , write two equations relating the first term of the series and the common difference.

    c. Hence find the first term and the common difference of the series .

    d. Find an expression for U_n

    e. Hence find U_{100}

16. Given that U_n = - 3n + 4 , Evaluate **a.** $\sum_1^{10} U_n$ **b.** $\sum_6^{12} U_n$ **c.** $10 + \sum_{10}^{30} U_n$

    d. $\sum_1^{30} U_n$ - $\sum_{10}^{30} U_n$ **e.** $10\sum_6^{12} U_n$

17. For a particular arithmetic series $\sum_6^{12} U_n$ = -5 and the first term is 0. Find the common difference and S_{20} .

18. For a particular arithmetic series $S_{10} = $ -5 and the first term is 0. Find the common difference and U_{10}

19. For an arithmetic series $U_5 = 7$ and $U_{15} = $ -3 .

a. Find an expression for U_n

b. Workout S_{15} and $\sum_5^{15} U_n$

c. Workout $\sum_5^n U_n$ in terms of n.

20. For a series , $S_n = pn^2 + qn$ also given $U_1 = 0$, $S_5 = 10$

a. Write 2 equations relating p and q

b. Hence find the values of p and q

c. workout $S_{n+1} - S_n$

d. Hence deduce U_n and justify that the series is arithmetic.

e. evaluate $\sum_5^{13} U_n$

21. For a series , $S_n = pn^2 + rn$ also given $U_1 = $ -5, and $S_2 = 4$

a. Write 2 equations relating p and r

b. Hence find the values of p and r

c. workout $S_{n+1} - S_n$

d. Hence deduce U_n and justify that the series is arithmetic.

e. evaluate $\sum_7^{13} U_n$

22. Given that $U_n = -3n + 5$,

a. Find S_n in terms of n

b. Evaluate $\sum_{10}^{101} U_n$

23. For a series $S_{k+1} - S_k = 5k - 7$

a. Find $S_{51} - S_{50}$ **b.** Find the 51st term   **c.** State U_k **d.** Hence workout $\sum_{4}^{12} U_k$

24. For a series $S_{k+2} - S_{k+1} = -k + 7$

a. Find the 27nd term   **b.** Find the 51st term   **c.** State U_k **d.** Hence workout $\sum_{k+4}^{K+12} U_k$

25. Given that for a particular $\sum_{n+1}^{n+2} U_n = 5n + 4$

a. State U_n

b. Justify that the series is arithmetic

c. Hence find S_{100}

26. A series given by   X , X -3 , x-6 , ... And the sum of of **n ,(n > 1)** terms is equal to 24

  a. state the common difference

  b. show that $3n^2 - (3+2x)n + 48 = 0$

  c. considering the fact that" n "has to be a **real number ,**Show that its necessary, $(3 + 2x)^2 \geq 576$

  d. solve the inequality $(3 + 2x)^2 \geq 576$ and find the range of x

  e. given that x = 24 find the value of **n**

27. A series given by $4, 4 + 2t, 4 + 4t, 4 + 6t$ .... And the sum of of **n** terms is equal to $\frac{25}{4}$

a. Show the series is Arithmetic .

b. show that $4tn^2 + (16\text{-}4t)n - 25 = 0$

c. considering the fact that" n "has to be a **real number** ,Show that its necessary, $t^2 + 17t + 16 \geq 0$

d. solve the inequality $t^2 + 17t + 16 \geq 0$ and find the range of t

e. explain why the condition $t^2 + 17t + 21 \geq 0$ is **necessary but not sufficient**.

28. Sum of n terms of an arithmetic series given by S_n , the n^{nt} term is given by u_n and the first term is **a** .

a. Write an expression for S_n in terms of **n** , **a** and u_n

b. Given that $S_{100} = 500$ and $u_{100} = 25$, find the **first term** and the **common difference** of the series

c. Hence write a **recurrence relationship** for this series .

29. The first term and the last term of an arithmetic series is given by -4 and 0 respectively . nth and the n+1th terms given by u_n and u_{n+1} respectively .

a. find a relationship between the sum of the series and the number of terms

b. Given that the total sum of the series is - 100 , find a recurrence relationship for this arithmetic series.

c. evaluate u_{10}

30. Sum of n terms of an arithmetic series given by S_n , the n^{nt} term is given by u_n and the first term is **a** .

a. Write an expression for S_n in terms of **n** , **a** and u_n

b. Given that $S_{101} = 0$ and $u_{101} = 25$, find the **first term** and the **common difference** of the series

c. Hence write a **recurrence relationship** for this series .

d. State u_n and Evaluate $\sum_5^{10} u_n$

31. For an arithmetic series $u_{11} + u_6 = 0$ and $u_3 + u_{21} = 14$

  a. Find u_1 and common difference

  b. Find an expression for u_n

  c. evaluate u_{21} and $\dfrac{u_3}{u_{21}}$

32. $\sum_3^7 u_n = 10$ and $\sum_5^{10} u_n = 0$, Given that u_n produces an arithmetic series find an expression for u_n . Hence find a recurrence relationship for this series.

32. K^{th} term (U_k) of an arithmetic series is " **a** " , the common difference by " **d**" and **nth** term is by U_n

  a. Show that for this series , $\sum_k^{2k} U_n = \dfrac{d}{2}k^2 + k(a + \dfrac{d}{2}) + a$

  b. Given that $\sum_k^{2k} U_n = k^2 - 20k - 21$,

    (i) State the common difference and the value of the kth term

    (ii) Complete the square of $k^2 - 20k - 21$ and find the **minimum** value of $\sum_k^{2k} U_n$

    (iii) Find the value of K where $\sum_k^{2k} U_n$ is minimum.

33. **Prove** that sum of an arithmetic series (s_n) is given by $\dfrac{n}{2}(2a + (n-1)d$, where **a** is the first term and **d** is the common difference , hence show that s_n can be re written as $\dfrac{d}{2}n^2 + n(a - \dfrac{d}{2})$

34. p^{th} term(U_p) of an arithmetic series is " **a** " , the common difference by " **d**" and n^{th} term is by U_n

a. Show that for this series , $\sum_{p}^{2p} U_n = \frac{d}{2} p^2 + p(a + \frac{d}{2}) + a$

b. Given that for an arithmetic series $\sum_{p}^{2p} U_n = 2p^2 - 50p + a$

(i) Find the value of a and the common difference

(ii) Write another expression for $\sum_{p}^{2p} U_n$ in terms of a , U_{2p} and p

(iii) by using parts " **b**" and (ii) Show that $U_{2p} = 4p - 52$

(iv) Find the value of p where $\sum_{p}^{2p} U_n = 0$

35. For a series $U_n = (-1)^n + 3n^2$, produce the first 5 terms of the series , given that $U_n = 299$, find the first 3 values of n, where $U_n > 100$

36. Given that, $U_n = (-1)^n \frac{n}{2n-10} + 1$,

a. produce the first 5 terms of the series

b. Given that $U_n = \frac{14}{9}$, find the value of n

c. Find **the range of values** of n, where $U_n < 2$

37. $U_n = 3n^2 - 60n - 63$,

a. By completing the square find the minimum value of U_n

b. Find the value of n where U_n is minimum

c. Find the range of values of n where $U_n \geq 0$

d. Hence state the first 3 vales of n which satisfies $U_n > 0$

38. Given that, $U_n = \frac{(-1)^n}{2n-10} + n$, produce the first 5 terms of the series , given that $U_n = \frac{7}{2}$, find the value of n , and find the range of **odd values** of n, where $U_n > \frac{-11}{12}$

39. An arithmetic sequence contain 50 terms, the sum of the terms of the **first half** of the sequence is twice as the sum of the terms of the **second half** of the sequence

a, Find an equation which relates the first term of the series and the common difference

b, Also given that the total sum of the series is 50, find the first term and the common difference

40. A man has a deposit (**D**) of a £10000 in a bank , He started to withdraw £200 each month from the 1st of every month from January 2014

a , How many months it will take him to withdraw all the money from the bank

b , Hence state the date he will withdraw is last £ 200 from the bank

c , find the date his deposit falls below £ 3000

d , Find the duration of time and dates the deposit (D) satisfies £2000 < **D** < £7000

41. A man had a deposit (**D**) of £50000 in a bank , also he was allowed to have an **overdraft** (drawing more money than the account holds) limit of £ 10000. He started to withdraw money from the 1st of each month from January the 2014 . His very first withdrawal was £ 400 and after each withdrawal was increased by £ 100 than the previous, and when he was withdrawing the money from the overdraft, his bank balance appeared as a negative value,

a , How many months it will take him to withdraw all his deposited the money from the bank

b , Find the duration of time the deposit (D) satisfies £15000 < **D** < £47000

c , Calculate how long it would take to drop his bank balance below £ - 4000

42. A recurrence relationship given as, $U_n = kU_{n+1} - U_{n+2}$, $U_1 = 2$, $U_2 = -2$

a, Find U_3 and U_4 in terms of **k**

b, Given that K $\neq$ 0 , and $U_4 = 2$, find **k**

c, Hence evaluate $\sum_1^5 U_n$

43, A recurrence relationship given as, $U_n = kU_{n+1} - U_{n+2}$, $U_1 = 2$, $U_2 = -2$, $k \neq 0$

a, Find U_3 and U_4 in terms of **k**

b, Given that $\sum_1^4 U_n = 0$, find K

c, Hence evaluate $\sum_2^6 U_n$

44. A recurrence relationship given as, $U_n = kU_{n+1} - 10$, $U_1 = 2$, $U_2 + U_3 = 0$

a, Find U_2 and U_3 in terms of **k**

b, Hence find the value of **k**

c, State the first 5 terms of the series

d, Find the value of U_n when $\frac{U_{n+1}}{U_n} = 1$,

45. A recurrence relationship given as $U_n = \frac{PU_{n+1}}{3 - U_{n+2}}$, $U_1 = 2$, $U_2 = -3$

a, Find U_3 and U_4 in terms of **p**

b, Given that $U_4 = \frac{5}{2}$ find the value of **p**

c, State the first 5 terms of the series